I_{DDQ} TESTING OF VLSI CIRCUITS

edited by

Ravi K. Gulati
Ford Microelectronics, Inc.

and

Charles F. Hawkins
University of New Mexico

A Special Issue of Journal of Electronic Testing: Theory and Applications

Reprinted from Journal of Electronic Testing:
Theory and Applications
Vol. 3, No.4

Springer-Science+Business Media, LLC

FRONTIERS IN ELECTRONIC TESTING

A KLUWER INTERNATIONAL BOOK SERIES

Consulting Editor
Vishwani Agrawal

JOURNAL OF ELECTRONIC TESTING:
THEORY AND APPLICATIONS
Vol. 3, No. 4, December 1992

I_{DDQ} TESTING OF VLSI CIRCUITS

A Special Issue of
Journal of Electronic Testing

Library of Congress Cataloging-in-Publication Data

IDDQ testing of VLSI circuits / edited by Ravi K. Gulati and Charles F. Hawkins.
p. cm. -- (Frontiers in electronic testing series)
DDQ is subscript in IDDQ title; testing method also known as quiescent current testing.
Published also as vol. 3, no. 4 of Journal of electronic testing.
Includes index.
ISBN 978-1-4613-6377-4 ISBN 978-1-4615-3146-3 (eBook)
DOI 10.1007/978-1-4615-3146-3
1. Integrated circuits--Very large scale integration--Testing.
2. Metal oxide semiconductors, Complementary--Testing. I. Gulati, Ravi K. II. Hawkins, Charles F. III. Series.
TK7874.I3223 1993
621.39'5--dc20 92-39926
CIP

Fifth Printing 1997.
Originally published by Kluwer Academic Publishers in 1993
Softcover reprint of the hardcover 1st edition 1993

Printed on acid-free paper.

JOURNAL OF ELECTRONIC TESTING: Theory and Applications, 3, 287 (1992)

Foreword

Like many others, I was amazed by Mark Levi's presentation, *CMOS is most testable*, at the 1981 International Test Conference. In a CMOS circuit, excepting the brief period when the actual switching of gates takes place, no current is meant to exist between the power supply and ground. In practice, however, a very small residual current remains after the gates have acquired their steady state. If we find that there is more than the normal amount of current after the switching transients have subsided, then we should suspect a problem in the circuit. Using this observation, Levi devised tests for various faults like shorted and open transistors and bridging of normally unconnected signals. This method of testing has become known as I_{DDQ} or *quiescent current testing*. Since the current through all devices of a VLSI chip must pass through the common supply terminals, its measurement provides enormous observability. Many other types of failures, like delay faults and time-dependent failures that are caused by imperfect switching characteristic of transistors, can also increase the quiescent current. The current test can be an early indicator of poor reliability of a semiconductor device.

In the conventional style of testing, one attempts to propagate the logic states of the suspected nodes to primary outputs. This is done for all or most nodes of the circuit. For sequential circuits, in particular, the complexity of finding suitable tests is very high. In comparison, the I_{DDQ} test does not observe the logic states, but measures the integrated current that leaks through all gates. In other words, it is like measuring a patient's temperature to determine the state of health.

Despite the perceived advantages, during the years that followed its initial announcements, skepticism about the practicality of I_{DDQ} testing prevailed. The idea, however, provided a great opportunity to researchers. New results on test generation, fault simulation, design for testability, built-in self-test, and diagnosis for this style of testing have since been reported. After a decade of research, we are definitely closer to practice. Yet no textbook on the subject has appeared.

In early 1991, the *Journal of Electronic Testing: Theory and Applications* (*JETTA*) published a special issue on boundary scan. Considering the success of that special issue, the editorial board of the journal voted to initiate a *Frontiers in Electronic Testing Series*. The topics for this series are to be selected based upon timeliness, perceived potential for being an emerging technology and the lack of available collected material. Some volumes in the series may be developed as special issues of *JETTA* and republished as stand-alone books. I am thankful to Carl Harris for making the series a reality.

I_{DDQ} testing certainly meets the above criteria. I believe the present volume will be useful to practitioners and researchers alike. In selecting the material for this volume, the editors, Ravi Gulati and Charles Hawkins, have accomplished a difficult job. On one hand, the articles must meet the journal's peer review and selection process. On the other hand, to make the volume self contained, one must strike a balance between introductory and advanced material. They deserve all the credit for their remarkable achievement.

I_{DDQ} Testing is the first volume in the *Frontiers in Electronic Testing Series*.

Vishwani D. Agrawal
Editor-in-Chief
AT&T Bell Laboratories
va@research.att.com

JOURNAL OF ELECTRONIC TESTING: Theory and Applications, 3, 289 (1992)

Preface

Power supply current monitoring to detect CMOS IC defects during production testing quietly grew its roots in the mid-1970s. Both Sandia Labs and RCA in the United States and Philips Labs in the Netherlands practiced this procedure on their CMOS ICs. At that time, this practice stemmed simply from an intuitive sense that CMOS ICs showing abnormal quiescent power supply current (I_{DDQ}) contained defects. Later, this intuition was supported by data and analysis in the 1980s by Levi (RADC), Malaiya and Su (SUNY—Binghamton), Soden and Hawkins (Sandia Labs and The University of New Mexico), Jacomino and co-workers (Laboratoire d'Automatique de Grenoble), and Maly and co-workers (Carnegie Mellon University).

Interest in I_{DDQ} testing has advanced beyond the data reported in the 1980s and is now focused on applications and evaluations involving larger volumes of ICs to improve quality beyond what can be achieved by previous conventional means. There is growing interest in algorithms necessary to support automatic test generation (ATG) for I_{DDQ} testing as well as for diagnosis of defects using I_{DDQ} test information. Built-In Current (BIC) monitors have been proposed as a more efficient alternative for I_{DDQ} measurements. This special issue on I_{DDQ} testing is therefore timely and includes papers that reflect all of these significant trends.

JETTA posted an open call for papers for this issue. The papers were selected through the peer review process of *JETTA*. In addition to reviewer acceptance, we emphasized papers with data from real manufacturing processes or that used silicon test circuits to verify the result. Each paper, after receiving reviewer comments, underwent a minimum of three editorial revisions in an effort to make this a special issue of higher quality. Unfortunately, several of the submitted papers could not be included. We, however, thank all authors who submitted papers.

This issue includes 11 papers. Soden et al. review I_{DDQ} testing in the first paper and place it in recent context. The next three papers present independent evaluations of I_{DDQ} and compare it against conventional methods that use functional test sets and/or scan-based test vectors. Maxwell et al. from Hewlett Packard compare test results on 26,415 die that were subjected to scan path stuck-at fault (SAF) testing, speed testing with functional vectors, and I_{DDQ} testing. Perry shows the impact of I_{DDQ} testing on quality levels over a five year period on a manufacturing process at Storage Technology that had previously used 99.6% SAF coverage for all ASICs. McEuen from Ford Microelectronics shows reliability improvements and impact on burn-in associated with I_{DDQ} testing in a study with 59,024 ICs. All three papers have a common conclusion: defect levels and reliability are markedly improved with I_{DDQ} testing. The fifth paper, by Segura et al. presents new data on the electronic properties of gate oxide shorts, bridging, and open circuit defects in CMOS circuits.

The next two papers address the issue of obtaining test patterns for I_{DDQ} testing. Mao and Gulati describe their QUIETEST methodology for selecting I_{DDQ} test vectors from existing logic test patterns. This procedure has attracted attention because of its ability to select a small number of test vectors for I_{DDQ} testing without sacrificing coverage for modeled defects that would have been achieved by I_{DDQ} measurements on all of the test vectors. Chen and Abraham describe their algorithm for specifically generating I_{DDQ} test patterns to cover modeled defects by I_{DDQ} measurements in their integrated logic and I_{DDQ} test generation approach.

Defect diagnosis using I_{DDQ} test information is addressed in the next two papers. Aitken uses both I_{DDQ} failure information and logic values for diagnosis of leakage faults. Chakravarty and Liu present algorithms for diagnosis of bridging faults.

In the final two papers, Rubio and Figueras show a simple BIC monitor design and give results from silicon fabrication, and Maly and Patyra review BIC designs and provide BIC design rules. BIC monitors were first proposed for IC testing in 1988 and remain a strong possible solution for testing and diagnosis in high density electronic applications using I_{DDQ} failure information.

We would like to thank the reviewers for their careful work and the authors for their wonderful patience in responding to our suggestions for the multiple revisions of their papers.

Ravi K. Gulati
Ford Microelectronics, Inc.
uranus!gulati@uunet.uu.net

Charles F. Hawkins
The University of New Mexico and Sandia National Labs
hawkins@houdini.eece.unm.edu

JOURNAL OF ELECTRONIC TESTING: Theory and Applications, 3, 291-303 (1992)

I_{DDQ} Testing: A Review

JERRY M. SODEN
Sandia National Laboratories, Department 2275, PO Box 5800, Albuquerque, NM 87185-5800

CHARLES F. HAWKINS
The University of New Mexico, Electrical and Computer Engr. Dept., Albuquerque, NM 87131

RAVI K. GULATI AND WEIWEI MAO
Ford Microelectronics, Inc., 9965 Federal Drive, Colorado Springs, CO 80921-3698

Received August 24, 1992.
Editor: V.D. Agrawal

Abstract. Quiescent power supply current (I_{DDQ}) testing of CMOS integrated circuits is a technique for production quality and reliability improvement, design validation, and failure analysis. It has been used for many years by a few companies and is now receiving wider acceptance as an industry tool. This article begins with a brief history of CMOS ICs to provide perspective on the origin of I_{DDQ} testing. Next, the use of I_{DDQ} testing for IC quality improvement through increased defect and fault detection is described. Then implementation issues are considered, including test pattern generation software, hardware instrumentation, limit setting, IC design guidelines, and defect diagnosis. An extended reference list is provided to help the reader obtain more information on specific aspects.

Keywords: CMOS IC, I_{DDQ}, Current testing, IC quality, defects, fault models.

1. Introduction

It is natural to be skeptical when someone says it is possible to determine the state-of-health of a complex integrated circuit with a few simple tests. Scientists and engineers are trained to be judgmental of what appear to be paranormal claims. Justifiable doubt in the unfamiliar can be aided by experience and investment in a traditional approach, such as stuck-at fault testing. What is often overlooked in the case of quiescent power supply current (I_{DDQ}) testing is that CMOS ICs are intrinsically designed for current testability. So what may appear to some to be black magic is in reality a tool that is inherently there to be used. This tool is presently being used to great advantage by some manufacturers and can be readily implemented by anyone willing to consider improved methods.

I_{DDQ} testing can significantly improve the quality and decrease the production cost of CMOS ICs. It is being merged successfully with conventional approaches, such as functional or stuck-at fault testing, to improve the detection of defects and failure mechanisms early in the manufacturing process. As an important part of a quality manufacturing approach, it provides information required for process evaluation metrics and corrective action. When combined with optical and electron beam diagnostic techniques, it is also useful for failure mechanism identification and defect localization (Figure 1 shows a CMOS IC open circuit detected with I_{DDQ} testing). In addition, I_{DDQ} testing is very effective for design verification and first silicon characterization, reducing time to market.

2. History

I_{DDQ} testing is linked to the history of CMOS IC design and fabrication. In 1963, a few months after starting work for Fairchild Semiconductor, Frank Wanlass originated and published the concept of complementary-MOS (CMOS) logic circuitry [1]–[2]. It occurred to him that a CMOS circuit *would use very little power* and that *in standby, it would draw practically nothing—just the leakage current*. The term *standby* here refers

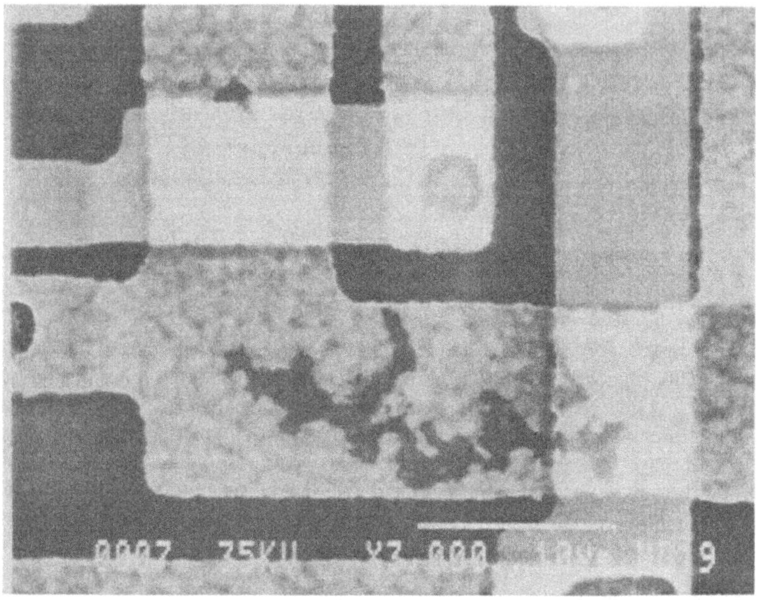

Fig. 1. CMOS IC open circuit detected with I_{DDQ} testing.

to the state when the inputs (and outputs) of the circuit are in a steady state condition. The first demonstration circuit, a two transistor inverter, consumed just a few nanowatts of standby power. Equivalent bipolar and PMOS gates consumed milliwatts of power even when in standby mode. CMOS shrank the standby power consumption by six orders of magnitude. It was therefore a fact that CMOS circuits with standby power consumption greater than a few nanowatts were *defective*.

Wanlass proved the CMOS concept using discrete p-channel and n-channel transistors. During the 1963–1964 time period, Israel Kalish led a group at RCA that built the first CMOS logic gates on a single chip [3]. Product families of small scale logic gates and simple sequential functions were introduced by RCA in 1968. CMOS ICs rapidly gained market share in the 1970s, riding the increasing tide of MOS microprocessor and memory sales for calculators and other applications. Digital watches were the first high volume CMOS product, an obvious application for low power consumption ICs. The first single chip CMOS microprocessor, RCA's 8-bit CDP1802 introduced in 1974, could be used easily in low power systems with battery back-up [3]–[5]. These CMOS ICs used Wanlass' fully static, fully complementary logic circuitry, thereby preserving the advantage of nanowatt standby power dissipation and I_{DDQ} testability.

I_{DDQ} testing is performed by measuring the current of the V_{DD} power supply in the quiescent logic condition after the inputs have changed (and prior to the next input change). Figure 2 shows the basic principles of I_{DDQ} testing. In this figure, a generalized IC has a gate

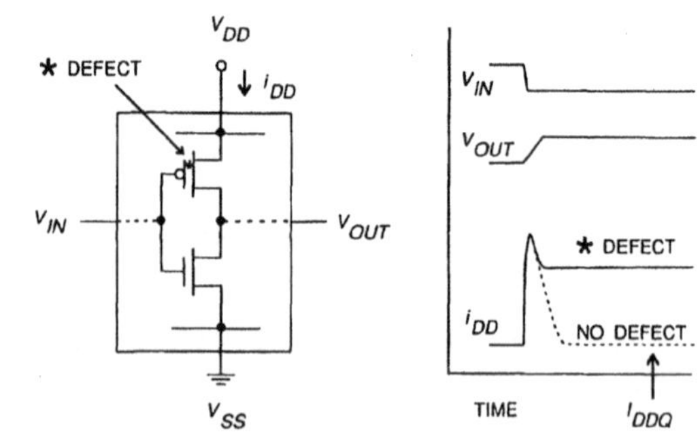

Fig. 2. Generalized CMOS IC with voltage and current waveforms.

oxide defect in a p-channel transistor of an internal logic gate. When the input to the logic gate changes from a logic 1 to a logic 0, the magnitude of I_{DDQ} changes from a low value to an elevated value. If the defect did not exist, I_{DDQ} would have returned to the initial low value. The vertical arrow on the time axis indicates when the I_{DDQ} measurement is made and emphasizes that some time is needed for i_{DD} to stabilize.

I_{DDQ} is the IEEE symbol for the quiescent power supply current in MOS circuits. In addition to *standby current* and *leakage current*, I_{DDQ} has been referred to with other terms and symbols such as power-down current, static current, monitored current, I_{ddq}, *IDDq*, *IDD*-All, *QIDD*, and I_{CC}. Some of these terms are incorrectly used. I_{DDQ} is the recommended symbol.

To assure low standby power consumption, some CMOS IC manufacturers included I_{DDQ} testing with their other traditional DC-parametric tests. They used several I_{DDQ} tests for SSI ICs and a somewhat higher number of tests for MSI and LSI ICs. For example, RCA performed I_{DDQ} tests at two power supply voltages and one input condition (such as several inputs low and the other inputs high) on their SSI ICs [6]. For their MSI and LSI ICs, they used the functional test to clock the IC into various logic states and then performed an I_{DDQ} test. Typical 25°C specifications for I_{DDQ} ranged from 1 to 500 nA. Manufacturers often observed a strong correlation between high I_{DDQ} and functional or dynamic test failure [7]. Nelson and Boggs found a correlation between elevated I_{DDQ} and increased propagation delay [8]. Using a special tester they found that I_{DDQ} tests could be used to detect pattern sensitive failures caused by increased access times in memories [9].

In the mid-1970s, some manufacturers, such as Philips in the Netherlands and Sandia National Labs [10] in the United States, began to use extended I_{DDQ} testing to improve defect detection and enhance the reliability of their ICs. For the relatively simple ICs manufactured during this time, I_{DDQ} tests performed at all functional test vectors constituted a check of all possible logic gate states for the IC, producing a very thorough test for defects and failure mechanisms. These I_{DDQ} tests at each functional test vector were performed slowly, using the precision measurement unit or simple custom circuitry on testers such as the Sentry from Fairchild [10]. However, the total I_{DDQ} test time was acceptable for low volume, high reliability ICs because the number of functional test vectors was usually less than ten thousand. In 1983 Toshio Maruyama designed a faster power supply current measurement circuit for test equipment being used by a CMOS watch IC manufacturer. This feature, now identified as the "bit current" option for Advantest testers, was the first capability specifically marketed for I_{DDQ} testing by a commercial test equipment company.

As functional complexity of CMOS ICs increased, the number of test vectors increased and it became impossible to achieve 100% gate-level logic state test coverage. I_{DDQ} tests were therefore performed on a portion of the functional vectors, often selected manually by the designer or test engineer using their knowledge of the IC. High nodal *activity*, sometimes based on logic simulation node toggling statistics, was usually the desired goal. Research in the mid-1980s on the effects of common CMOS IC defects, such as gate oxide shorts, improved the selection of test vectors for defect detection [11], [12]. Researchers also considered the use of I_{DDQ} test patterns for stuck-at, stuck-open, and delay fault coverage [13]–[15]. Improved methods for I_{DDQ} test pattern selection have been developed recently. A recent tool called QUIETEST has demonstrated that the selection of a small subset of I_{DDQ} test vectors within a large functional test set can provide the same leakage fault coverage as obtained by I_{DDQ} testing for the entire functional test set [16].

Competitiveness in CMOS IC manufacturing in the 1990s requires significant, continuous improvement in quality, reliability, and fabrication yields. Companies with long range goals for success are implementing comprehensive quality management programs that assure the development of quality processes for all work, from product conception through design, fabrication, testing. delivery, and customer usage. Considerable impact can be made towards achieving higher quality by incorporating I_{DDQ} testing along with structured design for testability and conventional logic testing. Companies have successfully implemented I_{DDQ} testing in various ways. Recent examples include using it to reduce manufacturing costs by over 50%, to significantly improve assembly line and field quality beyond that achieved by conventional scan design testing with higher than 99% stuck-at fault (SAF) coverage, and to cut burn-in failures in half [17]–[20].

3. Defect Detection

Difficulties with process capabilities critical to CMOS IC manufacturing, such as good coverage of oxide steps by metallization, have routinely plagued the industry [21] and intelligent decisions for process improvement require specific information on defects and failure mechanisms. I_{DDQ} testing is a very sensitive method for identifying defects and failure mechanisms [12], [22]. As a pass/fail production test, I_{DDQ} testing provides a better quality metric for physical defects, as opposed to conventional testing based solely on the unrealistic assumption that all IC abnormalities produce internal stuck-at faults or incorrect logic outputs. For failure analysis, I_{DDQ} techniques can provide rapid failure mechanism identification and physical localization. These include the use of I_{DDQ} vs. V_{DD} signatures, analysis of the relationship between I_{DDQ} test vectors and the logic structure of the IC, and digital processing of electron beam voltage contrast images [11], [23]–[28]. Also, optical measurements of light emission from the IC for high I_{DDQ} logic states can be used for defect localization [29].

I_{DDQ} testing for identifying common CMOS IC defects and failure mechanisms during manufacturing is relatively well understood. Typical defects include gate oxide shorts, interconnect bridging shorts, and interconnect open circuits and their effects upon logic behavior are often complex and time variant. Physical shorts and their related fault models bring to mind an image of a circuit schematic with two nodes bridged by a low value *ideal* resistor (e.g., $< 50\ \Omega$). In reality, shorts in ICs often have resistances greater than 1 kΩ or have nonlinear characteristics. Interconnect bridging shorts have been shown to have node to node resistances ranging from several hundred ohms to greater than 20 kΩ due to the geometry or contact characteristics of the bridging material [30]–[34]. Gate oxide shorts commonly exhibit diode behavior due to the polysilicon and substrate silicon doping differences [11], [23], [35]–[37].

Open circuits may not behave as ideal circuit discontinuities. IC interconnect *open* circuits have been found to have unstable, high resistances at contact interfaces [38] and nonlinear electrical characteristics [39]. Open circuit defects have been found to produce static node voltages that may turn on p- and n-channel transistor pairs, thus elevating I_{DDQ}.

Defects and failure mechanisms often initially cause increased I_{DDQ} without logical malfunction (functional failure). It may seem costly to reject ICs that have increased quiescent power supply current but are still performing their logic functions. However, it has been demonstrated that the defects and mechanisms responsible for elevated I_{DDQ} often greatly reduce reliability [19], [20], [35], [40], [41]. Also, high I_{DDQ} can cause premature failure of battery (power-limited) systems. In addition, there are quality and yield concerns. High I_{DDQ} can be symptomatic of a significant processing problem that may not be in control. Defects and failure mechanisms may worsen in time, resulting in major yield or reliability problems. Ignoring these defects and mechanisms can produce a false sense of security due to inaccurate quality or performance metrics.

4. Fault Detection

Practitioners of I_{DDQ} testing have used it primarily to target specific CMOS defects, but recently interest has focused on the effectiveness of I_{DDQ} to detect faults. Since faults are an abstraction, I_{DDQ} testing for faults targets those defects that cause well-defined fault behavior. If I_{DDQ} testing detects all defects that are modelled as a certain fault behavior, then I_{DDQ} testing is said to have achieved 100% coverage for that fault type. In the following, we examine the relationship of I_{DDQ} and fault detection because of the traditional interest in faults and the fact that fault models are critical for efficient I_{DDQ} automatic test generation (ATG) algorithms.

Faults are technology dependent [42], [43]. For example, logic gate bridging faults have different behavioral properties in TTL (wired-AND), ECL (wired-OR), and CMOS (wired-X), and the fault models take these into account [22], [44]. Similarly, the properties of the CMOS stuck-open fault are not found in other technologies [45], [46]. Faults are only useful if linked to real physical defects. The I_{DDQ}, or current sensing, test approach contrasts with conventional fault detection methods that use voltage sensing. For example, a logic gate bridging fault may be an absolute short (e.g., $< 50\ \Omega$) connection between the outputs of two logic gates. The voltage sensing technique first applies a test vector that activates the fault by placing opposite logic values across the fault, then sensitizes two paths from the outputs of the two logic gates to primary outputs where the voltage levels are read for a fault detection inference. The same bridging fault can be detected in CMOS ICs with current sensing if the test vector activates the fault. The I_{DDQ} test does not require the sensitized paths to primary outputs and therefore ATG requirements are significantly reduced. We will examine the various popular fault models and summarize the detection of each fault type by I_{DDQ} detection.

Stuck-at Faults. The stuck-at fault (SAF) model has dominated test methods since its origins in bipolar technology. The CMOS technology inherited the SAF model as a primary test metric with little experimental basis. Even today, the SAF link to defects and CMOS manufacturing test processes has not been observed in a significant percentage of defects [14], [47], [48]. Two articles in this volume (Maxwell *et al.* and Perry) and the work by Storey *et al.* illustrate the weakness of the sole use of the

SAF model in modern CMOS IC testing [18], [49]–[51]. Despite this, SAF coverage remains an established metric for communication between suppliers and customers.

I_{DDQ} testing can be linked to the SAF model if we examine all types of defects that occur in CMOS ICs. Levi observed that I_{DDQ} could detect the bridging of a logic node to V_{DD} or V_{SS} [13]. Malaiya and Su showed a relationship between I_{DDQ} testing and SAF behavior [14]. For bridging of logic nodes to V_{DD} or V_{SS}, I_{DDQ} testing will detect SAF behavior if the test vectors attempt to drive all logic gate I/O nodes to logic 0 and 1 [15]. Gate oxide shorts can also cause SAF behavior if low ohmic connections (on the order of 1 kΩ to 5 kΩ) are formed between the transistor gate and source diffusion region. This is a nondominant form of the gate oxide short that has been found in test structures [11] and is detectable by I_{DDQ} testing. In general, most bridging defects that cause SAF behavior are detectable with I_{DDQ} testing.

Many types of defects that cause open circuits elevate I_{DDQ}. The relationship of open circuits to elevated I_{DDQ} and SAF behavior needs careful examination. It is sometimes assumed that an open in the pullup (or pulldown) path will cause the logic gate output node to be stuck at V_{SS} (or V_{DD}). This behavior may not always be observable because the open causes a stuck-open fault rather than a SAF [45], [46]. Open circuit defects occurring in the gates of individual transistors have also been suggested as causing SAFs under the assumption that the transistor is off. This is generally not a correct assumption [52], [53] and, if it were, the defect again causes a stuck-open fault. A stuck-open (-off) fault is not a stuck-at fault and vice-versa.

Open circuit defects that occur on the inputs to logic gates (not individual transistors in the logic gate), as in Figure 3, cause a variety of electrical responses. If the break is small (on the order of a few hundred Å or less), then signal coupling can occur by electron tunneling [39]. The resultant increase in signal rise or fall time causes a delay fault and delayed elevation of I_{DDQ}. However, if the open defect is large, then a variety of responses are possible. Apparent SAF behavior may be observed with some responses causing detectable I_{DDQ} elevation and other responses causing zero or undetectable I_{DDQ} elevation. Logic gate inputs with large open defects are usually in a high impedance or floating node state. The node voltage (V_{fn} in Figure 3) seeks a steady state value that is a function of the local circuit parasitic elements. When V_{fn} *sticks* at a steady state value then the output voltage correspondingly exhibits logical stuck behavior and this output value can be a weak or strong logic voltage because of the analog gain of the logic gate. A weak output voltage occurs when V_{fn} acquires a value that is greater than the threshold voltages of both p- and n-channel transistors (i.e., 0.7 V < V_{fn} < 4.3 V for $|V_{TN}| = |V_{TP}| = 0.7$ V and $V_{DD} = 5$ V). If V_{fn} acquires a voltage that is less than either the p- or n-channel transistor threshold, then apparent SAF behavior occurs that may not be detectable by elevated I_{DDQ}. This type of defective circuit was reported for a test structure in which the floating input was attached to a bonding pad and the large metal capacitance pulled the input voltage to the substrate ground [39]. This type of defect is probable, but its occurrence for an actual CMOS IC has not been reported.

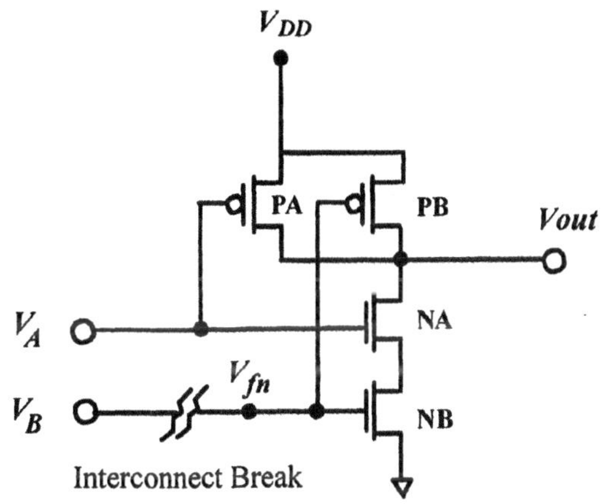

Fig. 3. 2-NAND open circuit defect.

In summary, defects that cause true SAF behavior in CMOS ICs are a small percentage of the real defects. Apparent SAF behavior can occur, but its dependence upon specific IC design and layout features invalidates the application of a generic, technology independent SAF model. SAF behavior due to defects that cause bridging to the power rails or cause both n- and p-channel pairs to be simultaneously on can be detected in most cases by I_{DDQ} tests.

Bridging Faults. Bridging faults can occur either at the logic output of a gate or at transistor nodes internal to the logic gate. I_{DDQ} testing for detection of logic node bridging faults originated with Levi [13] and was reported in practice by Acken [54]. Storey and Maly reported I_{DDQ} bridging fault detection and compared it with logic-based bridging fault detection [55]. Detection of 100% of these faults is possible when I_{DDQ} is measured with opposite voltage levels across each of the bridged nodes. I_{DDQ} testing has an advantage in that it can also detect resistive bridging defects that may cause delay faults, but not necessarily cause functional failure at lower test frequencies.

Bridging faults at the transistor level were analyzed by Jacomino *et al.* who examined the relation between I_{DDQ} testing and stuck-on faults [56]. Mao and Gulati used a software tool called QUIETEST to select I_{DDQ} test vectors to cover all six transistor node bridge combinations between gate, drain, source, and bulk [57]. This approach detects all transistor stuck-on faults and includes important defects such as gate oxide shorts, drain to source punch through, and leaky pn junctions between source, drain, and bulk.

CMOS Stuck-Open Faults. The stuck-open fault causes a high impedance state in a logic gate output node and under certain topological or design situations, I_{DDQ} is elevated and the fault is detected. I_{DDQ} testing does not guarantee detection, but has detected stuck-open faults in manufacturing test operations [46]. Stuck-open faults associated with one of the transistors in a CMOS transmission gate will elevate I_{DDQ} due to the body effect mechanism [22], [56]. Mao and Gulati's method selects I_{DDQ} vectors for the general case of transmission gates driving multi-input logic gates [57].

Delay Faults. Studies of the electrical properties of defects have shown that most of the random CMOS defects cause a timing delay effect rather than a catastrophic failure [22], [56], [58]. Most CMOS IC defects elevate I_{DDQ}. Any defect in a circuit signal path that elevates I_{DDQ} will finitely alter the rise and fall times of that signal. Thus, many delay faults are detectable with small numbers of I_{DDQ} test vectors. However, there is a subset of defects that cause delay faults but do not elevate I_{DDQ}. Such defects may cause increased interconnect resistance, particularly at vias, or they may cause an increase in transistor threshold voltage [58]. For these delay faults, it is relatively difficult for test generation as they require a sequence of two vectors for detection [59].

Use of I_{DDQ} for detection of delay faults was recently reported by workers at Motorola [34]. They found that metal bridging shorts in 256K SRAMs caused data path timing delay to increase by 3 ns and I_{DDQ} to increase by a factor of 2 to 3. Delay faults are now more of a concern since high speed CMOS products with clock rates in excess of 100 MHz are being designed. These shorter 5–10 ns clock periods make the ICs much more sensitive to circuit imperfections that are more difficult to detect. I_{DDQ} testing offers an inexpensive solution for detection of many CMOS IC delay faults.

In summary, I_{DDQ} plays a complementary role with logic response tests in the detection of faults [18], [22], [50]–[51], [58], [60]. The detection of SAFs with I_{DDQ} is a moot issue since SAFs have not been reported to be a concern for CMOS ICs. I_{DDQ} test patterns for logic gate bridges have been shown to be efficient compared to SAF test patterns, but data have not been reported for large circuits. I_{DDQ} test patterns are highly recommended for delay faults despite the lack of total delay fault detection capability. I_{DDQ} testing for the various fault models has strong advantages in vector count reduction, less restrictive controllability requirements, and increased observability of difficult faults.

5. Test Patterns

An I_{DDQ} test set uses vectors that create one or more low resistance paths from V_{DD} to V_{SS} in the presence of defects in CMOS circuits. By definition, if the circuit under test contains defects detectable by an I_{DDQ} test vector, then the circuit draws excessive quiescent current upon the application of that test vector.

There are three types of I_{DDQ} test patterns. *Every Vector* I_{DDQ} test patterns are those where the entire logic test pattern set is used and I_{DDQ} is measured for each test vector. In *Selective* I_{DDQ} test patterns, a subset of the logic test vectors is selected for I_{DDQ} measurements. Test vectors can also be generated specifically for I_{DDQ} testing; they comprise *Supplemental* I_{DDQ} test pattern sets. While *Every Vector* I_{DDQ} test patterns have several advantages, the number of test vectors for which I_{DDQ} must be measured is usually too large for the methodology to be practical in large volume production testing of VLSI chips.

When *Selective* I_{DDQ} test patterns are used, logic testing is carried out at normal speed except for occasional pauses for I_{DDQ} measurement when a selected I_{DDQ} test vector is applied. A methodology for obtaining *Selective* I_{DDQ} test patterns was first proposed by Mao *et al.* [16]. Their QUIETEST system selects I_{DDQ} test vectors to cover gate oxide shorts and transistor stuck-on faults (leakage faults). The results show that often less than 1% of the logic test vectors could be selected for I_{DDQ} testing to provide as much leakage fault coverage as would be obtained by *Every Vector* I_{DDQ} testing. This was a significant result that made I_{DDQ} testing practical in large volume production test environments. The QUIETEST system has since been expanded by Gulati *et al.* [61] to cover weak faults.

The weak fault model represents defects that cause degradation in logic levels without affecting logic values. Vandris and Sobelman [62] have also proposed a mixed functional/I_{DDQ} testing methodology for CMOS transistor faults. *Selective* I_{DDQ} tests are employed to cover transistor stuck-on faults that cannot be detected by logic testing.

In *Supplemental* I_{DDQ} testing, logic testing is done at full system speed and is supplemented by I_{DDQ} measurements for vectors generated specifically for I_{DDQ} testing. Obviously, the testing proceeds at a slower clock rate for the I_{DDQ} vectors. Bollinger and Midkiff [63] proposed a test generation methodology for bridging faults in combinational static CMOS logic. Chen and Abraham have proposed an approach for generating tests for switch level circuits using both I_{DDQ} and logic test generation; node stuck-at, transistor stuck-open, and transistor stuck-on faults are considered [64], [65].

At present, the most practical approach recommended for industrial use is *Selective* I_{DDQ} testing. It has been effective for several fault types on 40,000 transistor sequential circuits and has economical production test times. As more effective *Supplemental* I_{DDQ} test generation approaches for sequential circuits are developed, *Selective* I_{DDQ} testing may be replaced or augmented with *Supplemental* I_{DDQ} vectors.

6. Instrumentation

I_{DDQ} measurement requires analog circuitry that can ideally measure current below 1 μA at rates above tens of kHz. These goals are not easy to obtain and compromises are often practiced. The I_{DDQ} measurements in production testing use off-chip instrumentation with the sensor and its circuitry placed either on the tester load board or inside the tester itself. Early measurement circuits used diode comparator circuits, operational amplifiers, or simply a current viewing resistor connected in series with the I_{DDQ} path to the V_{DD} pin [10], [12]. These circuits typically achieve I_{DDQ} limit measurements on the order of 10–20 μA, but recent data show that these I_{DDQ} levels are well above serious defect detection in CMOS ICs. More sensitive circuits use the Keating and Meyer floating pin technique [66], [67], precision circuitry in the tester power supply [68], or built-in current (BIC) monitors in which the I_{DDQ} detection is done on-chip [69]–[74].

Several factors can degrade off-chip I_{DDQ} instrumentation. These include: (1) the pulse width of the CMOS IC transient current, (2) the impedance loading of the tester, (3) current leakages into or out of the tester, and (4) the high noise environment of the load board [12], [68]. All CMOS circuits have a relatively large transient current associated with switching of the logic gates. These amplitudes can vary from a few mA to several amperes depending upon the circuit design and the strength and loading of the output drivers. Signal and clock input timing skews can also widen the transient pulse width, further degrading the measurement sensitivity. All testers have capacitive loading at their probes that can vary from 20–200 pF. This affects I_{DDQ} measurement since the V_{DD} pin sees this impedance when the output driver is high and sees virtually nothing when the output driver is low. Thus, the variable load impedance on the V_{DD} pin is a function of the output driver logic state. This introduces considerable noise into the measurement [12]. The problem can be eliminated if the output pins are physically disconnected or tri-stated during the test and if they are not multiplexed I/O pins. I_{DDQ} testing of ICs with multiplexed I/O pins generally presents a noisier environment requiring a slower test rate. Tester current leakage presents a similar logic state dependent noise to the V_{DD} pin of the IC. The load board exists in a noisy electrical environment; long leads are used and electromagnetic fields are high. One proposed design integrates the I_{DDQ} monitor into the power supply in an attempt to reduce these problems and to improve software control [68].

A circuit proposed by Keating and Meyer [66] has been used by several workers because it has good sensitivity (sub μA) and its measurement speed at the 1 μA level can be tens of kHz [67]. Figure 4(a) illustrates an approach for the Keating and Meyer off-chip I_{DDQ} instrumentation and Figure 4(b) shows a typical current waveform response when a defect is present. The decay curve may follow an RC decay as shown or could be a constant current source response (constant slope). The response curve for short time intervals is given by

$$i_{DD} = \frac{C\Delta V}{\Delta t} \qquad (1)$$

The tester measures ΔV and Δt during the quiescent time and can estimate I_{DDQ} if C is known. C is the total capacitance at the V_{DD} node and includes that due to the IC itself and any capacitance added by the tester and instrumentation circuit. C typically has values on the order of several nanofarads for VLSI circuits and picofarads for smaller SSI circuits. For SSI or MSI

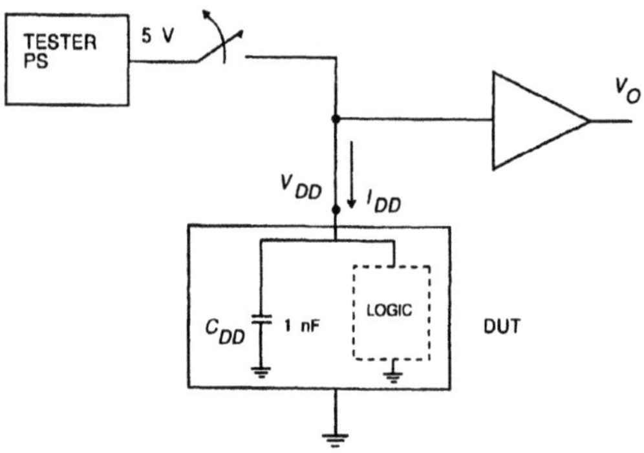

Fig. 4a. Off-chip instrumentation schematic.

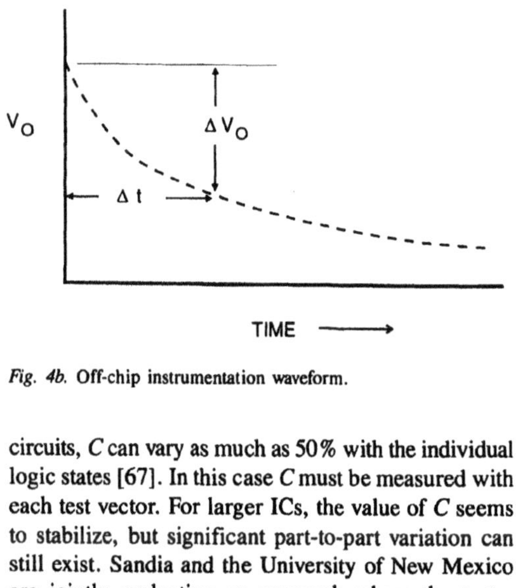

Fig. 4b. Off-chip instrumentation waveform.

circuits, C can vary as much as 50% with the individual logic states [67]. In this case C must be measured with each test vector. For larger ICs, the value of C seems to stabilize, but significant part-to-part variation can still exist. Sandia and the University of New Mexico are jointly evaluating an approach where the tester measures C. Off-chip I_{DDQ} measurements have been made above a 100 kHz rate for a 1 μA limit. Further evaluation to determine its production environment capability continues.

A recent idea uses built-in current (BIC) monitors as part of the IC circuitry to allow a more sensitive, noise-free measurement [69]–[74]. On-chip BIC monitors are reported to operate much faster than off-chip instrumentation. A BIC monitor operating at a 2 MHz test rate with a 2 μA sensitivity was developed by workers at Philips [73]. The transient pulse widths seen by the BIC monitor circuit are shorter since the contribution from the output buffers may be avoided by separating the BIC circuit from the output buffer ground returns. The BIC circuit is usually placed in the source lines of the NMOS transistors of the normal logic. This can significantly degrade performance of the IC during normal operation unless the design includes separate ground pins for the normal operation and for BIC monitor testing. With separate ground pins, the normal operation ground pin bypasses the BIC monitor for normal mode and is open circuited during BIC monitor mode. With separate ground pins, there is no IC speed degradation due to the BIC monitor. Maly and Patyra discuss BIC monitor design rules in [74].

BIC monitors have not been reported for production IC testing, but their biggest use in the future may lie in their value as diagnostic tools in high density electronic systems. Commercial testers are beginning to offer off-chip I_{DDQ} instrumentation. Improvements in I_{DDQ} off-chip instrumentation sensitivity, test measurement rate, and human friendliness are expected in the future.

7. Limit Setting, Characterization, and Reliability

Production I_{DDQ} testing requires selection of a pass/fail value for the current limit. Experience shows that this limit should be selected on a case-by-case basis because there is no single value that is generically right for all situations. The selection depends upon the manufacturer's specific design, fabrication and test environment, and the user's requirements. Proper selection requires evaluation of test data from representative ICs. Some vendors perform limited I_{DDQ} testing, such as a single I_{DDQ} measurement per IC, and these data may be used to help set the initial limit for multiple I_{DDQ} tests. Even when data are available, it may still be appropriate to perform a thorough characterization of samples of ICs from recent wafer lots. For these small quantities of ICs, it is practical to characterize I_{DDQ} using every vector from the functional test program and to measure I_{DDQ} with a slow, precision measurement unit. These data can then be statistically analyzed [12].

I_{DDQ} histograms can be analyzed by examining the relative scatter of the data for various test vectors. An example is shown in Figure 5. This figure shows that the majority of the ICs evaluated have low, defect-free current, but there is a distribution of high current devices ranging from a few to several hundred microamperes. A single failure mechanism, gate oxide shorts, was responsible for the high current. The impact of setting a limit in the several μA range (< 10 μA) can be easily determined from this figure. The effect of solving the gate oxide problem can also be readily evaluated from such data.

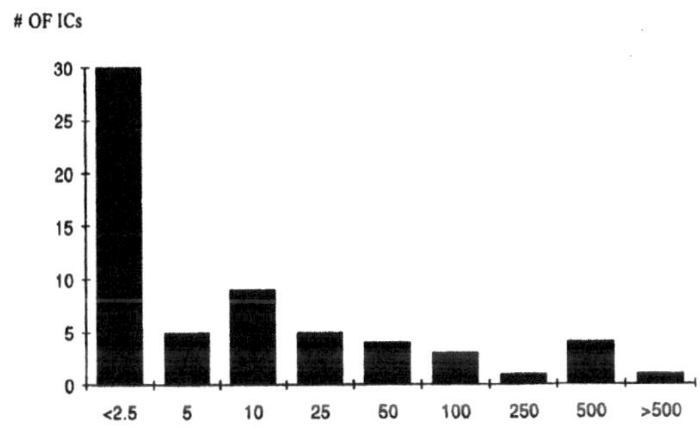

Fig. 5. Histogram of I_{DDQ} for a 32 bit microprocessor.

Data such as those in Figure 5 enable quick determination of the potential values for the I_{DDQ} test limit and the effect this limit will have upon yield. These data also help identify candidate defects for corrective action. For example, there may be an obvious multimodal distribution with sufficient separation among the regions to clearly indicate the priorities for corrective action. The percentage of ICs with high I_{DDQ} provides an estimate of the reliability risk that exists if they are not removed. Agressive manufacturing should and can drive the I_{DDQ} limit to < 1 μA. I_{DDQ} data often expose many previously undetected (or ignored) defects and mechanisms [17]–[20], [35].

The I_{DDQ} test limit directly influences the choice of production test instrumentation. Test circuits have been evaluated for pass/fail testing and for parametric (analog) testing situations. Typical production test circuits have *acceptable* test vector rates for I_{DDQ} test limits in the 1 to 20 μA range. *Acceptable* is emphasized because it is often assumed that I_{DDQ} testing will be an additional test. It is commonly considered unacceptable to increase test time unless strong quality or reliability benefits can be demonstrated. Our experience shows that the incremental I_{DDQ} test time is justifiable for the production of high quality, high reliability ICs.

For companies using commercial ICs, the vendor's data book I_{DDQ} specification can be a good initial limit for improved detection of bridging defects and other defects that cause relatively high current [51]. Simple tests with only a few vectors and a relatively high 0.5–1 mA limit have been shown to be effective for detecting defects not caught by conventional scan design testing even with 99.6% stuck-at fault coverage [17]. A limit of 70 nA has been used for 1.5 micron technology SRAMs [23]. Limits in the 1 μA range have been shown to be effective for state-of-the-art ICs, such as sea-of-gate ASICs with three metallization layers and 0.55 micron effective channel lengths [60].

I_{DDQ} testing can detect reliability-limiting failure mechanisms at wafer probe testing, enabling the reduction of the burn-in time normally required to accelerate these mechanisms to the point of functional failure [19]–[20]. To achieve high reliability, I_{DDQ} testing can either be targeted for specific, known defects or combined with functional, stuck-at, or delay fault testing vectors.

8. IC Design

As discussed in Section 1, ICs designed to the original definition of CMOS logic (100% static and complementary) have nanowatt power dissipation in the quiescent state [1]. As a result, completely CMOS ICs are inherently designed to be I_{DDQ} testable with no additional circuitry necessary. We have also shown that some ICs that are not fully CMOS logic designs can still be tested using I_{DDQ} measurements [75]. For example, 80C51 8 bit microcomputers with dynamic precharge circuitry were given I_{DDQ} tests at 50 kHz using 20,000 vectors from the vendor's test program. About 300 design-related high current states were found and the test vectors for these were masked out of the test set. These high current states in commercial CMOS ICs can occur when the main objective of the IC designer is to assure that proper logic functions are achieved and maintenance of fully CMOS logic is a secondary concern. For example, microprocessors may have logic states during which the internal data bus is not used. In these states, the bus may not be driven or may be driven simultaneously by more than one driver. These are logical *don't care* states for the bus, but the floating inputs of the data bus receivers or the contention of the multiple drivers will cause high I_{DDQ}.

As another example, memories may have pullup resistors instead of p-channel transistors in the array. These ICs may be described in their data sheets as completely static CMOS, but I_{DDQ} will be milliamperes when the IC is enabled due the noncomplementary design of the memory cell. Despite designed-in high current due to pull-up resistors, I_{DDQ} testing has been used to identify shorts that caused timing delays in 256K SRAMs [34]. ICs with intentional or unintentional high current conditions can result from design practices that originated with higher power dissipation technologies such as TTL and NMOS. It is common to hear that designers have included pull-up resistors in input buffer cells simply because they became accustomed to using these resistors for earlier IC technologies or PC board designs and did not consider other alternatives more appropriate for CMOS ICs.

It is also common to hear the issue of silicon area overhead mentioned as a reason for not designing fully CMOS circuits. Experience at Sandia Labs has shown that overhead penalties, if they exist, are outweighed by the benefits of fully CMOS designs. For example, workers at Sandia recently emulated National Semiconductor's 32C016 16/32 bit NMOS microprocessor in a fully CMOS version designated the SA3300. The SA3300 has about 65,000 transistors, nearly the same number as in the 32C016. Because of the SA3300's fully CMOS design, high SAF coverage can be achieved using less than 3000 I_{DDQ} test vectors, an accomplishment that would be impossible using conventional techniques for the NMOS design. For the commercial 80C51 mentioned previously, Sandia workers found they could achieve very high SAF coverage for the fully CMOS version using I_{DDQ} testing with the vendor's test set, whereas conventional fault simulation of the commercial version was estimated to take over 21 years using a MACH 1000 hardware accelerator.

The microprocessor and memory examples in this section and the preceding section demonstrate that the capabilities of I_{DDQ} testing do not diminish significantly as the IC complexity increases. The number of transistors increases as the IC complexity grows, but the size of the transistors and their junctions usually decreases with technology advancement. Thus the background junction leakage current has remained substantially unchanged with higher levels of integration.

It is important to ensure that the logic is implemented in a manner that maintains the low power dissipation capability of CMOS. This requires that the targets and margins for parameters such as thin and thick oxide threshold voltage be appropriately chosen. If, for example, the thin oxide threshold voltage drops to about 0.5 V or less, it is common for conventional technologies that subthreshold leakage may begin to exceed the reverse-biased junction leakage. As the magnitude of the background leakage increases, it becomes increasingly difficult to detect design and defect problems.

In addition to greatly increased defect and fault coverage, fully CMOS designs, such as the SA3300, enable the use of current testing for immediate identification of design, layout, and fabrication problems. In several cases, companies have been convinced of the value of I_{DDQ} testing first through its use for design validation/verification rather than for defect detection. For example, one gate array company uses I_{DDQ} testing routinely to identify routing errors that leave logic gate inputs unconnected. The unconnected, floating inputs cause elevated I_{DDQ} that is readily detected with a few test vectors.

9. Defect Diagnosis

The objective of defect diagnosis is to make use of the I_{DDQ} test patterns to locate defects in ICs. Defect localization is one aspect of failure analysis. Defect diagnosis can be used to improve yields when a new design or process is introduced, to monitor processes for high volume production, or to identify weak points later in the life cycle by analysis of failed chips from system assembly or customers. Conventional techniques for location of resistive shorts include hot-spot detection, waveform probing, and emission microscopy [29].

I_{DDQ} testing is an effective alternative for location of resistive shorts and other defects in CMOS circuits. Diagnosis is possible by *Every Vector, Selective,* and *Supplemental* I_{DDQ} test patterns and practical systems have been implemented [16], [24]. Aitken [25]–[27] has proposed that since I_{DDQ} monitoring alone provides only one observation point, it may be advantageous to couple it with logic observations for more effective diagnosis. Chakravarty and Liu have created algorithms for diagnosis of bridging faults [76].

Selective I_{DDQ} test patterns are usually worse than *Every Vector* I_{DDQ} test patterns in their diagnosis capability. When *Selective* I_{DDQ} testing is used in production, *Every Vector* I_{DDQ} testing may still be employed for better fault diagnosis on failing chips. *Supplemental* I_{DDQ} test patterns can be better or worse in their diagnosis effectiveness compared to other types of I_{DDQ} test patterns. Combined use of logic observations and I_{DDQ} monitoring may improve the diagnosis capability.

10. Conclusion

I_{DDQ} testing is an effective technique for improving the quality and reliability of CMOS ICs. Successful methodologies exist for implementing this technique rapidly into high volume manufacturing processes. New software and hardware approaches have been proposed recently that expand the options available for incorporating I_{DDQ} testing into mainstream production. The recent achievements of this technique are impressive and include significantly reduced product cost and improved process yields.

Acknowledgment

We would like to thank all who have contributed to the field of I_{DDQ} testing. We also thank Vishwani D. Agrawal for his thorough review and valuable suggestions for this article. The work performed at Sandia National Laboratories was supported by the U.S. Department of Energy under contract DE-AC04-76DP00789.

References

1. F.M. Wanlass and C.T. Sah, "Nanowatt logic using field-effect metal-oxide semiconductor triodes," *Proc. Solid State Circuits Conf.*, pp. 32–33, February 1963.
2. M.J. Riezenman, "Wanlass's CMOS circuit," *IEEE Spectrum*, p. 44, May 1991.
3. *EE Times*, Issue 503A, September 1988.
4. R.N. Noyce and M.E. Hoff, Jr., "A history of microprocessor development at Intel," *IEEE Micro*, pp. 8–21, February 1991.
5. *User Manual, CDP1802 COSMAC Microprocessor*, RCA Corporation, November 1977.
6. J. Flood, "Fundamentals of testing COS/MOS integrated circuits," *RCA Solid-State Integrated Circuits Databook*, Application Note ICAN-6532, pp. 622–630, 1978.
7. R.G. Daniels, "The changing demands of microprocessor testing," Keynote address, *Proc. Int. Test Conf.*, September 1990.
8. G.F. Nelson and W.F. Boggs, "Parametric tests meet the challenge of high-density ICs," *Electronics*, pp. 108–111, December 11, 1975.
9. G.F. Nelson; personal communications, March 1988.
10. W.J. Barnard, G.W. Garrett, and L.K. Horning; personal communication, August 1992.
11. C.F. Hawkins and J.M. Soden, "Electrical characteristics and testing considerations for gate oxide shorts in CMOS ICs," *Proc. Int. Test Conf.*, pp. 544–555, November 1985.
12. C.F. Hawkins, J.M. Soden, R.R. Fritzemeier, and L.K. Horning, "Quiescent power supply current measurement for CMOS IC defect detection," *IEEE Trans. on Indus. Electron.*, vol. 36, no. 2, pp. 211–218, May 1989.
13. M. Levi, "CMOS is most testable," *Proc. Int. Test Conf.*, pp. 217–220, October 1981.
14. Y.K. Malaiya and S.Y.H. Su, "A new fault model and testing technique for CMOS devices," *Proc. Int. Test Conf.*, pp. 25–34, November 1982.
15. R.R. Fritzemeier, J.M. Soden, K.R. Treece, and C.F. Hawkins, "Increased CMOS IC stuck-at fault coverage with reduced I_{DDQ} test sets," *Proc. Int. Test Conf.*, pp. 427–435, September 1990.
16. W. Mao, R.K. Gulati, D.K. Goel, and M. Ciletti, "QUIETEST: A quiescent current testing methodology for detecting leakage faults," *Proc. ICCAD*, pp. 280–283, November 1990.
17. R.J. Perry, "I_{DDQ} testing in CMOS digital ASIC's—putting it all together," *Proc. Int. Test Conf.*, pp. 151–157, September 1992.
18. R.J. Perry, "I_{DDQ} testing in CMOS digital ASIC's," *J. of Electronic Testing: Theory and Applications (JETTA)*, vol. 3, no. 4, pp. 317–325, December 1992.
19. S.D. McEuen, "IDDq benefits," *Proc. IEEE VLSI Test Symp.*, pp. 285–290, April 1991.
20. S.D. McEuen, "Reliability benefits of I_{DDQ}," *J. of Electronic Testing: Theory and Applications (JETTA)*, vol. 3, no. 4, pp. 328–335, December 1992.
21. T. Frederiksen, *Intuitive IC CMOS Evolution*, Donnelley Press, 1984.
22. J.M. Soden and C.F. Hawkins, "Electrical properties and detection methods for CMOS IC defects," *Proc. 1st European Test Conf.*, pp. 159–167, April 1989.
23. C.T. Wang, "Voltage-current characteristic model for gate-oxide short induced standby current failures in CMOS ICs," *Proc. Electrical Overstress/Electrostatic Discharge Symp.*, pp. 246–251, September 1987.
24. D.J. Burns, "Locating high resistance shorts in CMOS circuits by analyzing supply current measurement vectors," *Proc. Inter. Symp. for Testing and Failure Analysis*, pp. 231–237, November 1989.
25. R.C. Aitken, "Fault location with current monitoring," *Proc. Int. Test Conf.*, pp. 623–632, October 1991.
26. R.C. Aitken, "A comparison of defect models for fault location with current monitoring," *Proc. Int. Test Conf.*, pp. 778–787, September 1992.
27. R.C. Aitken, "Diagnosis of leakage faults with I_{DDQ}," *J. of Electronic Testing: Theory and Applications (JETTA)*, vol. 3, no. 4, pp. 363–375, December 1992.
28. R. Bottini, D. Calvi, S. Gaviraghi, A. Haardt, D. Turrini,, and M. Vanzi, "Failure analysis of CMOS devices with anomalous IDD currents," *Int. Symp. for Testing and Failure Analysis (ISTFA)*, pp. 381–388, November 1991.
29. C.F. Hawkins, J.M. Soden, E.I. Cole, Jr., and E.S. Snyder, "The use of light emission in failure analysis of CMOS ICs," *Int. Symp. for Testing and Failure Analysis (ISTFA)*, pp. 55–67, October 29–November 2, 1990.
30. R. Rodriguez-Montanes, E. Bruls, and J. Figueras, "Bridging defects resistance measurements in a CMOS process," *Proc. Int. Test Conf.*, pp. 892–899, September 1992.
31. J.M. Acken, "Deriving accurate fault models," Ph.D. thesis, Technical Report No. CSL-TR-88-365, Stanford University, CA, October 1988.
32. W. Yarbrough, "A testing methodology and test chip design strategy for IC fabrication process assessment, problem diagnosis, and yield analysis," Ph.D. thesis, Stanford University, Stanford, CA, April 1988.
33. H. Hao and E.J. McCluskey, "On the modeling and testing of gate oxide shorts in CMOS logic gates," *IEEE Intl. Workshop on Defect and Fault Tolerance in VLSI Systems*, Hidden Valley, PA, November 1991.
34. T. Guckert, P. Schani, M. Phillips, M. Seeley, and N. Herr, "Design and process issues for elimination of device failures due to 'drooping' vias," *Proc. Int. Symp. for Testing and Failure Analysis (ISTFA)*, pp. 97–101, November 1991.
35. C.F. Hawkins and J.M. Soden, "Reliability and electrical properties of gate oxide shorts in CMOS ICs," *Proc. Int. Test Conf.*, pp. 443–451, September 1986.
36. H. Hao and E.J. McCluskey, "Resistive shorts within CMOS gates," *Proc. Int. Test Conf.*, pp. 292–301, October 1991.
37. J.A. Segura, V.H. Champac, R. Rodrigues-Montanes, J. Figueras, and J.A. Rubio, "Quiescent current analysis and experimentation of defective CMOS circuits," *J. of Electronic Testing: Theory and Applications (JETTA)*, vol. 3, no. 4, pp. 337–348, December 1992.
38. A.N. Campbell, E.I. Cole, C.L. Henderson, and M.R. Taylor, "Case history: failure analysis of a CMOS SRAM with an intermittent open contact," *Proc. Int. Symp. for Testing and Failure Analysis (ISTFA)*,, pp. 261–269, November 1991.

39. C.L. Henderson, J.M. Soden, and C.F. Hawkins, "The behavior and testing implications of CMOS IC logic gate open circuits," *Proc. Int. Test Conf.*, pp. 302–310, October 1991.
40. J.M. Soden and C.F. Hawkins, "Reliability of CMOS ICs with gate oxide shorts," *Semiconductor Int.*, pp. 240–245, May 1987.
41. C.W. Green, "PMOS dynamic RAM reliability—a case study," *Proc. Int. Reliability Physics Symp*, pp. 213–219, April 1979.
42. J.A. Abraham, "VLSI Testing," Chapter 1. *Fault Modeling in VLSI*, Amsterdam: North-Holland, Elsevier Science Publishers, pp. 1–27, 1986.
43. S. Gai, M. Mezzalama,, and P. Prinetto, "A review of fault models for LSI/VLSI devices," *Software & Microsystems*, vol. 2, pp. 44–53, April 1983.
44. S.D. Millman and J.P. Garvey, "An accurate bridging fault test pattern generator," *Proc. Int. Test Conf.*, pp. 411–418, October 1991.
45. R.L. Wadsack, "Fault modeling and logic simulation of CMOS and MOS integrated circuits," *Bell Systems Tech. J.*, vol. 57, pp. 1449–1448, May–June 1978.
46. J.M. Soden, R.K. Treece, M.R. Taylor, and C.F. Hawkins, "CMOS IC stuck-open fault electrical effects and design considerations," *Proc. Int. Test Conf.*, pp. 423–430, August 1989.
47. F.J. Ferguson and J.P. Shen, "Extraction and simulation of realistic CMOS faults using inductive fault analysis," *Proc. Int. Test Conf.* pp. 475–484, September 1988.
48. R.R. Fritzemeier, C.F. Hawkins, and J.M. Soden, "CMOS IC fault models, physical defect coverage, and I_{DDQ} testing," *Proc. Custom Integ. Circuits Conf. (CICC)*, pp. 13.1.1–13.1.8, May 1991.
49. P.M. Maxwell, R.C. Aitken, V. Johansen, and I. Chiang, "The effectiveness of I_{DDQ}, functional, and scan tests: how many fault coverages do we need," *Proc. Int. Test Conf.*, pp. 168–177, September 1992.
50. P.M. Maxwell and R.C. Aitken, "I_{DDQ} testing as a component of a test suite: the need for several fault coverages," *J. of Electronic Testing: Theory and Applications (JETTA)*, vol. 3, no. 4, pp. 305–316, December 1992.
51. T. Storey, W. Maly, J. Andrews, and M. Miske, "Stuck fault and current testing comparison using CMOS chip test," *Int. Test Conf.*, pp. 311–318, October 1991.
52. M. Renovell and G. Cambon, "Topology dependence of floating gate faults in MOS circuits," *Electr. Letters*, vol. 22, pp. 152–153, January 1986.
53. W. Maly, P.K. Nag, and P. Nigh, "Testing oriented analysis of CMOS ICs with opens," *Proc. Int. Conf. Comp. Aid. Design (ICCAD)*, pp. 344–347, November 1988.
54. J.M. Acken, "Testing for bridging faults (shorts) in CMOS circuits," *Des. Auto. Conf.*, pp. 717–718, June 1983.
55. T. Storey and W. Maly, "CMOS bridging fault detecting," *Proc. Int. Test Conf.*, pp. 842–851,, September 1990.
56. M. Jacomino, J.L. Rainard, and R. David, "Fault detection by consumption measurement in CMOS circuits," *Proc. 3rd Int. Conf. on Fault Tol. Comput. Sys.*, pp. 83–94, F. Belli and W. Dorke (ed.), Berlin: Springer-Verlag, 1987.
57. W. Mao and R.K. Gulati, "QUIETEST: a methodology for selecting I_{DDQ} test vectors," *J. Electronic Testing: Theory and Applications (JETTA)*, vol. 3, no. 4, pp. 349–357, December 1992.
58. J.M. Soden, R.R. Fritzemeier, and C.F. Hawkins, "Zero defects or zero stuck-at faults—CMOS IC process improvement with I_{DDQ}," *Int. Test Conf.*, pp. 255–256, September 1990.
59. J.L. Carter, V.S. Iyengar, and B.K. Rosen, "Efficient test coverage for delay faults," *Proc. Int. Test Conf.*, pp. 418–427, September 1987.
60. K. Sawada and S. Kayano, "An evaluation of I_{DDQ} versus conventional testing for CMOS sea-of-gate ICs," *Proc. Int. Test Conf.*, pp. 158–167, September 1992.
61. R.K. Gulati, W. Mao, and D.K. Goel, "Detection of 'undetectable' faults using I_{DDQ} testing," *Proc. Int. Test Conf.*, pp. 770–777, September 1992.
62. E. Vandris and G. Sobelman, "A mixed functional/I_{DDQ} testing methodology for CMOS transistor faults," *Proc. Int. Test Conf.*, pp. 608–614, October 1991.
63. S.W. Bollinger and S.F. Midkiff, "On test generation for I_{DDQ} of bridging faults," *Proc. Int. Test Conf.*, pp. 598–607, October 1991.
64. C.H. Chen and J.A. Abraham, "High quality tests for switch-level circuits using current and logic generation algorithms," *Proc. Int. Test Conf.*, pp. 615–622, October 1991.
65. C.H. Chen and J.A. Abraham, "Generation and evaluation of current and logic tests for switch-level sequential circuits," *J. of Electronic Testing: Theory and Applications (JETTA)*, vol. 3, no. 4, pp. 359–366, December 1992.
66. M. Keating and D. Meyer, "A new approach to dynamic IDD testing," *Proc. Inter. Test Conf.*, pp. 316–321, September 1987.
67. K.M. Wallquist, A.W. Righter, and C.F. Hawkins, 'Implementation of a voltage decay method for I_{DDQ} measurement on the HP82000," *Hewlett-Packard User Group Mtg.*, San Francisco, CA, June 1992.
68. C. Crapuchettes, "Testing CMOS IDD on large devices," *Proc. Inter. Test Conf.*, pp. 310–315, September 1987.
69. D. Feltham, P. Nigh, L.R. Carley, and W. Maly, "Current sensing for built-in testing of CMOS circuits," *Proc. Int. Conf. on Computer Des.*, pp. 454–457, Rye Brook, NY, October 1988.
70. W. Maly and M. Patyra, "Built-in current testing," *IEEE Trans. on Solid State Ckts.*, vol. 27, pp. 425–428, March 1992.
71. Y. Miura and K. Kinoshita, "Circuit design for built-in current testing," *Proc. Int. Test Conf.*, pp. 873–881, September 1992.
72. J. Rius and J. Figueras, "Proportional testing for current testing," *J. Electronic Testing: Theory and Applications (JETTA)*, vol. 3, no. 4, pp. 387–396, December 1992.
73. A. Welbers, B. Verhelst,, E. Seevinck, and K. Baker, "A built-in CMOS I_{dd} quiescent monitor circuit," Philips internal report, 1989.
74. W. Maly and M. Patyra, "Design of ICs applying built-in current testing," *J. of Electronic Testing: Theory and Applications (JETTA)*, vol. 3, no. 4, pp. 397–406, December 1992.
75. L.K. Horning, J.M. Soden, R.R. Fritzemeier, and C.F. Hawkins, "Measurements of quiescent power supply current for CMOS ICs in production testing," *Proc. Int. Test Conf.*, pp. 300–309, September 1987.
76. S. Chakravarty and M. Liu, "Algorithms for I_{DDQ} measurement based diagnosis of bridging faults," *J. of Electronic Testing: Theory and Applications (JETTA)*, vol. 3, no. 4, pp. 337–385, December 1992.

Jerry M. Soden received the B.S.E.E. from Kansas State University (1969) and the M.S.E.E. from the University of New Mexico (1978). He is a Senior Member of Technical Staff at Sandia National Laboratories in the Components Parts Center. Since 1976 his responsibilities have included IC failure analysis, reliability testing, and design for reliability and testability. Mr. Soden has 22 years of industry experience in IC quality and reliability and has authored papers on these topics, including several with outstanding paper awards. He is a member of the Technical Program Committee and Board of Directors of the Electrical Overstress/Electrostatic Discharge Symposium. He is a member of the 1992 Program Committee of the International Test Conference and the International Symposium for Testing and Failure Analysis.

Charles F. Hawkins received the B.S.E.E. from the University of Florida (1960), M.S.E.E. from Northeastern University (1964), and the Ph.D. from the University of Michigan (1970). He has worked with I_{DDQ} testing and CMOS IC defect electrical characterization since 1984. He is a Professor in the Electrical and Computer Engineering Dept. at the University of New Mexico and works as a contract consultant to Sandia National Labs Multichip and Custom Products department. He has authored papers on these topics, including several with outstanding paper awards, is active in teaching testability short courses for industry, and teaches VLSI Design and Test at the university. He is the Program Chair for the 1993 International Test Conference.

Ravi K. Gulati obtained his B.E. degree in Electrical and Electronics Engg. from B.I.T.S., Pilani, India in 1980. He worked as a Design Engineer at R&D Labs, Tata Electric Cos., Bombay from 1980 to 1983. He obtained the M.S. degree in Electrical and Computer Engg. from the University of Iowa, Iowa City in 1985. Since then Mr. Gulati is employed with Ford Microelectronics, Inc., at Colorado Springs, where presently he is Engineering Manager in the CAE & Advanced Technology organization.

Mr. Gulati has published and taught numerous tutorials on various aspects of testing including I_{DDQ}. He has served on the program committee of the 1991 IEEE VLSI Test Symposium. Presently, he is a program committee member of the International Test Conference and coordinates the area of CAE for defect detection and I_{DDQ} testing. Mr. Gulati is a member of the IEEE and its P1149.4 working group for the development of mixed-signal test bus standards.

Weiwei Mao received the B.S. and M.S. degrees in Electrical Engineering from Fudan University, Shanghai, People's Republic of China, in 1982 and 1984, and Ph.D. in Electrical Engineering from the University of Colorado, Colorado Springs, in 1991. He is presently working as a Principal Member in the CAE & Advanced Technology organization at Ford Microelectronics, Inc. in Colorado Springs. From 1986 to 1990, he was a Research Assistant in the Electrical and Computer Engineering department, University of Colorado at Colorado Springs. From 1984 to 1986, he was an Assistant Professor in the Electronic Engineering department at Fudan University.

Dr. Mao's research interests include I_{DDQ} testing, delay faults testing, automatic test generation, design for testability and testing of mixed signal circuits. He has widely published in the area of testing.

JOURNAL OF ELECTRONIC TESTING: Theory and Applications, 3, 305-316 (1992)

I_{DDQ} Testing as a Component of a Test Suite: The Need for Several Fault Coverage Metrics

PETER C. MAXWELL AND ROBERT C. AITKEN
Design Technology Center, Hewlett-Packard Company, Santa Clara, CA 95052

Received March 2, 1992. Revised July 30, 1992.
Editor: C.F. Hawkins

Abstract. This article is concerned with the role of I_{DDQ} testing, in conjunction with other types of tests, in achieving high quality. In particular, the argument is made that rather than use a single fault coverage, it is better to obtain a number of different coverages, for different types of faults. To demonstrate the need for increasingly stringent fault coverage requirements, an analysis is given of the relationship between quality, fault coverage and chip area. This analysis shows that as chip area increases, fault coverage must also increase to maintain constant quality levels. Data are then presented from a production part tested with I_{DDQ}, scan, timing and functional tests. To realistically fault grade I_{DDQ} tests, three different coverage metrics are considered. The data show differences in tester failures compared to these coverage metrics, depending on whether one uses total I_{DDQ} failures (parts which fail I_{DDQ} regardless of whether they fail other tests as well) or unique I_{DDQ} failures (parts which fail only I_{DDQ}). The relative effectiveness of the different components of the full test suite are analyzed and it is demonstrated that no component can be removed without suffering a reduction in quality.

Keywords: Current testing, fault coverage, functional testing, physical defects, quality, scan.

1. Introduction

To be able to assess the effectiveness of a test set in producing high quality parts, it is necessary to use some form of metric to grade the test set. Traditionally, the metric used is the single stuck-at fault coverage, even though it is acknowledged that this is a poor descriptor of faults in some types of circuits, notably CMOS [1]. Single stuck-at faults do affect the quality level of a part, but there are many other fault types possible (e.g., bridging, stuck-open, stuck-on, delay) in single and multiple occurrences. Quality level is therefore a function of *total* fault coverage. The single stuck-at coverage is used as an estimator, but the effectiveness of the estimation depends on the type of test set being graded. Ideally, the test set detects the same proportion of nonmodelled faults as it does of modelled faults. The estimate is not very accurate when the detection of these nonmodelled *peripheral* faults is relatively poor. Surprisingly, this can result in test sets with lower measured fault coverages actually producing better quality levels than other test sets with higher measured fault coverages. This was empirically demonstrated in a recent study [2] in which scan vectors with a coverage of 92% produced the same quality level as functional vectors with a coverage of 83%. Consequently, fault coverage goals need to be qualified with the type of test used.

Since the scan tests achieved good stuck-at detection and the functional vectors (apparently) achieved good detection of other faults, the combination of the two resulted in good quality levels for the part. This suggests a general test strategy which combines several different types of tests, each targeting different types of faults, and each graded with respect to those faults in order to obtain a better estimate of total fault coverage. However, the single stuck-at metric will probably continue to be used until more progress is made in targeting and grading other fault types, and until their effectiveness is demonstrated.

With this in mind, a study was done to extend the earlier results in two ways:

1. Extrapolation with chip size. The chip used for the experiment was small and had a high yield. The results could be used to help set fault coverage goals

for similar chips but it was desired to see how these goals would need to change for much larger designs.

2. Addition of extra tests. One method of explicitly targetting additional faults is to utilize current monitoring of the power supply, otherwise known as I_{DDQ} testing [3]. There have been studies reported on the effectiveness of this technique for specific failure types such as gate oxide shorts [4] and stuck-opens [5], as well as its impact on burn-in [6]. These studies do not concern themselves with the relative effectiveness of I_{DDQ} compared to other types of tests for general ASICs. This aspect has been addressed in simulation studies [7], [8] and empirically demonstrated for small chips [9], [10]. The purpose of the present study was to examine the performance of I_{DDQ} tests compared to scan and functional tests for a typical ASIC in normal production.

2. Quality Versus Chip Size

2.1. Quality Level and Faults

To understand the basis for the extrapolation presented here, an overview of the assumptions made is presented. Quality level will be expressed as *reject rate*, where

$$\text{reject rate} = \frac{\#\text{bad parts}}{\#\text{parts passing the test}}$$

This is sometimes referred to as *defect level* but this term is not used here to avoid any confusion with physical defects. Since quality levels and yields due to parametric and continuity failures are not subject to any form of analysis based on particle defects, this report considers the *functional* portion of the test only. In other words, yields are functional yields and not total yields.

Reject rates arise from bad parts being tested as good. If we denote the fault coverage by f, the probability of testing a bad part as good by $P_{bg}(f)$ and the yield as y, then the reject rate is:

$$RR = \frac{P_{bg}(f)}{y + P_{bg}(f)} \tag{1}$$

where the probability of testing a bad part as good is the sum:

$$P_{bg}(f) = \sum_{n=1}^{n=N} p(n) \times m(n) \tag{2}$$

where $p(n)$ is the probability of having n faults, $m(n)$ is the probability of the test missing n faults and N is the total number of all possible faults.

With a fault coverage of f, the probability of the test missing a single fault is $(1 - f)$. Assuming that the faults are independent, the probability of the test missing all n faults is $(1 - f)^n$. Thus we can write:

$$\begin{aligned} P_{bg}(f) = {} & \text{prob. of 1 fault} \times (1 - f) + \\ & \text{prob. of 2 faults} \times (1 - f)^2 + \\ & \text{prob. of 3 faults} \times (1 - f)^3 + \ldots \end{aligned} \tag{3}$$

For high fault coverages $(1 - f) << 1$, so that the major contributor to (3) is the first term, that is, the probability $p(1)$ that a chip will have a single fault. Even though the probabilities that a chip will have two or three faults are higher than $p(1)$, they are masked by the additional factors of $(1 - f)$. With very little loss of accuracy we can approximate reject rate as:

$$RR \approx \frac{P_{bg}(f)}{y} \tag{4}$$

We can then derive a simple expression for the probability of one fault:

$$p(1) \approx \frac{y \cdot RR}{1 - f} \tag{5}$$

To use (5), several estimates of variables are required. The reject rate is measured from returned parts and is a measure of real defects. Fault coverage f is total fault coverage, a figure which is not known. In practice this value is estimated by stuck-at fault coverage obtained from a simulator. Although this is less than desirable, the experimental results presented in [2] showed that for functional vectors, the stuck-at coverage was a reasonable estimate for total fault coverage. Finally, the value of y is obtained from the tester's goodpile/bonepile separation. The value is subject to error if the tester rejects good parts [11] and this error affects the measured accuracy of RR and y. We were aware of this problem existing but were unable to account for it in the results. We believe the error was minimized by analyzing all returned parts to determine the cause of failure, as described below.

Summarizing the data presented in [2], the functional yield was 0.767, the fault coverage was 0.989 and the observed final reject rate was 360 ppm. This reject rate was determined by the product division shipping back parts which failed during system test. From these returned parts it was possible to obtain the proportion that were true test escapes, as opposed to some other failure, such as bent pins, ball bond breakages or ESD damage. This was done by first verifying that the parts failed in a golden system, then verifying that they still passed the original IC test. The only portion of defective chips not known is that which has not yet failed in the field, but this number is assumed to be so low that it will not impact the results significantly. Substituting these values into the above equation for $p(1)$ gives a value of 0.025. To extrapolate predicted reject rates to chips of varying sizes, it is necessary to determine how $p(1)$ varies. This requires a study of the relationship between defects and faults.

2.2. *Defect Distributions and Faults*

We use the term *defect* for those physical occurrences which give rise to some circuit malfunction. Other physical occurrences are possible (landing in nonactive portions of the chip), but since they do not contribute to any yield loss they are not considered here. A *fault* is the logical manifestation of a defect.

Clearly, observed faults are produced from physical defects, but the distribution of physical defects over a number of chips is not the same as the distribution of the faults they cause. This is because one defect, on average, causes more than one fault and/or fault type. It will always cause at least one fault because we are considering as defects only those physical occurrences which give rise to faults. For example, a defect causing a pinhole short through the gate oxide of a p-transistor of a NOR gate could cause an input to output bridging fault, a slow-to-rise fault and excessive I_{DDQ}. Given this interpretation of a defect, the probability of zero defects is the same as the probability of zero faults (this value is the yield, Y). However, the probability of n defects is not the same as the probability of n faults. In particular, the probability of *one* defect is much more than the probability of one fault. This causes the fault distribution to be skewed in such a way that no single distribution can represent it. Fortunately, in our case we are interested in the probability of *one* fault, not in the fault distribution.

To obtain the ratio of probabilities of one defect and one fault, a defect distribution is required. A number of different distributions have been used for yield modelling purposes [12]. A very common yield model is Murphy's Equation [13], which relates the yield (probability of zero defects) to the product defect density × area. Unfortunately, it is difficult to generate a complete defect distribution using this model, so for the purposes of obtaining some actual numbers, we will use a negative binomial distribution [14], which takes defect clustering into account. The value of the clustering parameter α has been observed to typically lie between 0.3 and 5 [15]. We will use the value 4, which gives a yield very close to that predicted by Murphy's Equation. For a particular process, if the true value of α is known it could be used instead, but the results are not sensitive to the value of α. Thus, our defect distribution is given by:

$$\text{Probability of } n \text{ defects} = \frac{\Gamma(n+4)\left(\frac{DA}{4}\right)^{n}}{n!\,\Gamma(4)\left(1+\frac{DA}{4}\right)^{n+4}} \tag{6}$$

where $\Gamma(x)$ is the gamma function, D is the defect density and A is the effective area of the chip. From this equation, the yield is given by:

$$Y = \left(1+\frac{DA}{4}\right)^{-4} \tag{7}$$

The observed functional yield of 0.767 corresponds to a value of 0.28 for the DA product. Substituting this value into (6) and evaluating for different n results in Table 1. The left column gives the number of defects on a die and the right columm the probability of this number of defects actually occurring.

Table 1. Defect distribution.

# Defects	Probability
0	0.768
1	0.197
2	0.031
3	0.004

From this table we see that the proportion of die with one defect is 0.2, whereas the proportion of die with one fault, as calculated above, is only 0.025. Thus, whatever mechanism produces faults from defects, on average there are only about 1/8 as many die with one fault as there are with one defect. Since there is no reason to expect this to change with increasing area, it can serve as the basis for extrapolation. Note that this ratio would change if a different defect distribution were used, but the trends shown in the extrapolation below will remain.

2.3. *Extrapolation to Different Chip Sizes*

Rather than use chip area directly, the *DA* product is used. The extrapolation is based on the assumption that the ratio of the probabilities of 1 fault to 1 defect remains constant. The exact procedure is then:

1. For each *DA* value calculate the defect distribution.
2. Calculate the value of $p(1)$ as $\frac{0.025}{0.2}$ times the probability of 1 defect.
3. Calculate the reject rate.

The results are summarized in Tables 2-4 and in Figures 1-3. Table 2 (Figure 1) shows the variation of functional yield and the parameter $p(1)$ for a range of *DA* products. The data point for the chip discussed above is shown highlighted. It is interesting to observe that $p(1)$ increases at first, but then starts decreasing as *DA* increases. This is because larger chips have more faults on average and therefore the probability of a single fault is smaller. For small chips $p(1)$ decreases because the yield increases and the probability of there being any defects at all decreases.

Table 2. Variation of yield and $p(1)$ with area.

DA	0.25	**0.28**	0.5	1.0	1.5	2.0	2.5
Y	0.79	**0.77**	0.62	0.41	0.28	0.20	0.14
$p(1)$	0.023	**0.025**	0.035	0.041	0.038	0.033	0.028

Table 3. Variation of reject rate (ppm) for a constant fault coverage.

DA	0.25	0.28	0.5	1.0	1.5	2.0	2.5
Reject rate for 98.9% coverage	310	360	560	990	1380	1700	1970

Table 4. Required fault coverages for 1000 ppm and 500 ppm.

DA	0.25	0.28	0.5	1.0	1.5	2.0	2.5
Fault coverage for 1000 ppm	96.6	96.9	98.2	99.0	99.3	99.4	99.5
Fault coverage for 500 ppm	98.3	98.5	99.1	99.5	99.6	99.7	99.8

These values have been used in two ways. Table 3 (Figure 2) shows the resultant reject rates for a constant fault coverage of 98.9%. Table 4 (Figure 3) shows the required fault coverage to achieve reject rates of 1000 ppm and 500 ppm. To illustrate the relative insensitivity to the value of the clustering parameter α, we regenerated these tables using a value $\alpha = 1$. This corresponds to significantly more clustering but results in the required coverages being only approximately 0.4% lower for the 1000 ppm curve and 0.2% lower for the 500 ppm curve.

These results show that as chip sizes increase, fault coverage has to improve even to maintain a given quality level. At these high levels of total coverage required, it is important to obtain a better estimate than single stuck-at coverage. For this reason a study was carried out using I_{DDQ} tests, with regard to both their effectiveness in rejecting defective parts and how they can be graded.

3. Fault Simulation for I_{DDQ}

3.1. *Current versus Voltage Sensing*

Grading vectors which will be used in conjunction with I_{DDQ} measurements presents different problems than

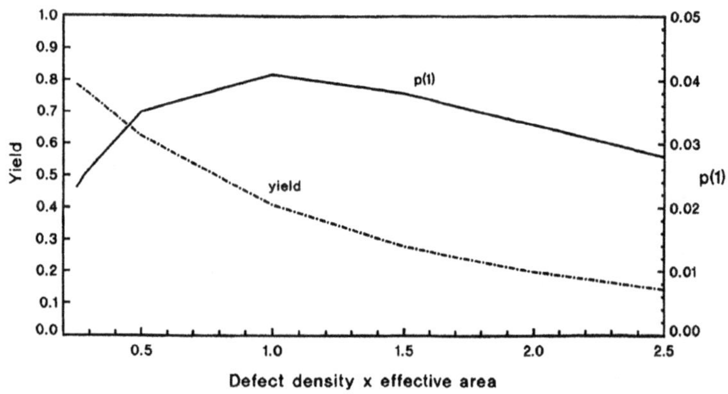

Fig. 1. Variation of yield and $p(1)$.

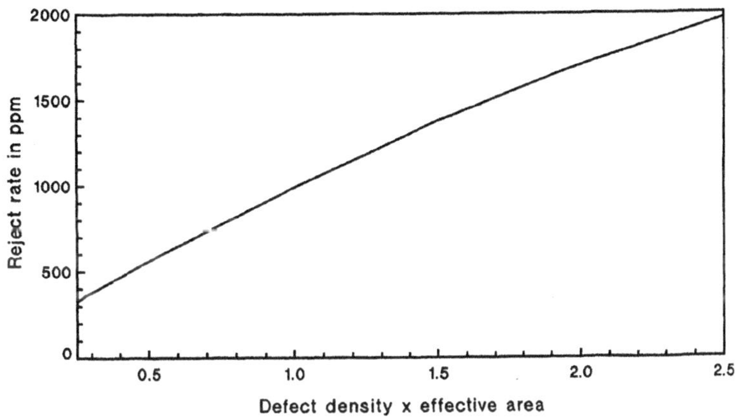

Fig. 2. Quality level obtained with 98.9% coverage.

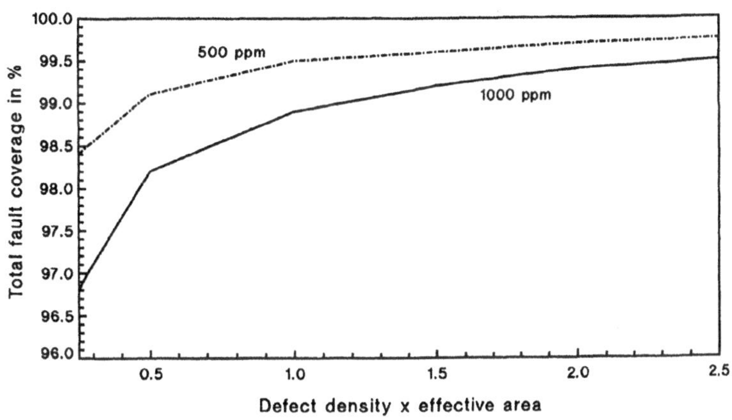

Fig. 3. Required fault coverages.

grading voltage sensing test vectors. The reason is that I_{DDQ} testing targets defects much more directly than voltage sensing. This means that a model such as the stuck-at model, which is a voltage model, is less appropriate. When I_{DDQ} is used, a stuck-at condition represents a ***true*** stuck-at, that is, a node shorted to V_{DD} or *GND*. For voltage sensing, other conditions can give rise to a node *behaving* as if it were stuck-at, but not necessarily giving rise to excessive current. An open is a defect which can cause this behavior under certain topological conditions. Experimental studies [16], [17] have shown that a floating gate of an inverter can result in the output voltage being V_{DD}, which corresponds to an input stuck-at 0 as far as voltage sensing is concerned. However, driving the input to logic 1 will simply not change the output level—it will not cause excessive current. Similar behavior can also result from an open in the pulldown structure. This results in a stuck-open fault condition, but can give stuck-at fault behavior if the gate is accessed frequently enough to prevent discharge of the output node.

The stuck-at coverage of an I_{DDQ} test set can be easily obtained [8] by determining if a node is driven to both 0 and 1 during the test, the so-called *toggle test.* However, since this type of fault is a small subset of possible faults, the coverage figure is of little use by itself in helping to determine quality levels. Consequently, a better model needs to be used.

Since I_{DDQ} is based on current measurements, an improved coverage metric is based on shorts, or bridges. This is based on the physical principle that current exists when two nodes are connected and driven to different values.

3.2. Pseudo-Stuck-at Faults

Based on a generalization of an argument in [18], if current monitoring is used, all shorts between the gate, source, drain, and substrate of transistors within a combinational CMOS gate are detected using I_{DDQ} on a conventional single stuck-at test set applied to the gate. To apply this result to complex gates, they must be modelled using AND, NAND, OR, NOR and NOT primitives, and the stuck-at faults must be propagated to the output of the actual gate, not the primitives. Consequently, a *voltage* stuck-at coverage is representative of the internal shorts coverage. This was used in [8] to obtain *stuck-at* coverage figures for a variety of ISCAS circuits. Conventional fault simulation must be modified for I_{DDQ} test sets by requiring each stuck-at fault to propagate through a single gate instead of to a primary output, since the observation is done via the power line. Single gate propagation ensures that a full stuck-at voltage test is applied to each gate.

The coverage figures obtained by such a process can be thought of as *pseudo-stuck-at* coverages, since the intent is to detect internal transistor shorts as well as hard stuck-ats. The advantage of using this metric is that there are minimal changes required to a conventional fault simulator to obtain the coverage.

3.3. Transistor Shorts

Pseudo-stuck-at coverage figures reflect the underlying model of transistor shorts. A more accurate coverage figure, therefore, is to consider these shorts explicitly, and report coverage figures in terms of them. The coverage figures we have obtained exclude a class of transistor shorts which may be characterized as *a priori* undetectable. These are shorts between transistor poles which are always connected, such as the source and substrate on an inverter, and can never be detected by any test. Conventional redundant faults, where the logical properties of the circuit prevent nodes from being driven to appropriate values, were not determined, so they appear as undetected in the coverage figures.

It is worth noting here that for both the transistor short case and the pseudo-stuck-at case the number of redundancies is typically much less than for voltage sensing. Normal logic redundancies arise when either a gate input combination cannot be realized (stimulation redundancy) or a fault effect cannot be propagated to a primary output (propagation redundancy). When using I_{DDQ}, only stimulation redundancies remain, since propagation is automatic using the power supply.

Obtaining coverage figures for these shorts requires the use of tables that indicate which shorts are detected for a given input pattern to a gate, plus a fault-free simulation of the circuit [19], [20].

3.4. Bridges

The above dealt with intra-gate bridges. Inter-gate bridges can also occur. We chose to model a set of realistic bridges, namely those occurring between adjacent metal lines. These were obtained from the circuit artwork using a modified capacitance extract function, in a manner similar to that of Carafe [21]. The extraction function essentially searched the routing channels for pairs of lines which were separated by the minimum metal pitch. The distance the lines were adjacent was also noted, which provides a method for comparing the likelihood of given defects. For this work, all extracted pairs were considered with the exception of those between V_{DD} and ground lines.

When I_{DDQ} measurements are used, a bridge may be said to be detected if the two nets which comprise it have opposite values in the fault-free circuit. Simulation is thus straightforward, as is ease of detection, since no values have to be propagated. This is a considerable improvement over voltage testing, since the actual logic levels on the nets (or portions of the nets, in the case of *resistive* bridges) do not have to be calculated, and oscillation is not a problem. In addition, resistive bridges that may not cause logic errors at low frequencies are detectable by I_{DDQ} [22].

While it may be argued that bridges between lines which are adjacent over a large portion of the chip are in some sense more likely to occur than those adjacent for only a few microns, the coverage figures we have obtained give all modelled defects equal weight.

3.5. Comparison of Different Metrics

To compare the above metrics, an I_{DDQ} test set was generated by modifying a conventional ATPG tool to use the pseudo-stuck-at model. All single stuck-at faults on gate inputs are targeted, and a fault is listed as detected if its effect propagates to its gate's output, rather than a primary output. Complex gates (including exclusive-ORs) are modelled by networks of simple gates and propagation in these cases must be to the complex gate output, rather than the output of a component *sub-gate*. Tests generated in this manner are sufficient to detect any single leakage failure within a cell. Multiple failing cells are also guaranteed to be detected, provided that there is only a single failure in each cell, and that the cells have independent input cones. In a typical scan circuit, this should account for a large fraction of potential multiple failures. It is also possible to take an existing set of scan vectors, fault simulate them, and retain only those vectors which add to pseudo stuck-at or transistor short coverage. Our experience indicates, however, that a shorter test set will result from directly generating I_{DDQ} tests.

Figure 4 shows the coverage profiles for the three models discussed. Given that pseudo-stuck-ats are intended to reflect transistor shorts it can be seen that the two do not track each other very well. In the early part of the test the pseudo-stuck-at coverage is considerably less than the transistor short coverage. Furthermore, the two never converge, with final figures being 94.5% and 98.7% respectively.

It is of interest to note that the coverages for transistor shorts and inter-gate bridges closely match, especially in the final values (98.7% and 98.5% respectively). For comparison purposes, coverage of the same inter-gate bridges using voltage sensing of the normal scan vectors is 84.1%.

4. The Performance of I_{DDQ}

4.1. Experimental Vehicle

Tests were carried out on a part which was the successor to the one in [2]. The part was a fully static standard cell design of 8577 gates (35904 transistors) and which had 436 flip-flops. The digital test suite consisted of a set of 59,000 functional vectors run at slow speed (2 MHz), a smaller subset run at high speed (20 MHz and 32 MHz), 357 scan tests and 141 I_{DDQ} tests. The 20 MHz tests were identical to a set run at 2 MHz while the 32 MHz tests were a subset of other 2 MHz tests. The stuck-at fault coverage of these tests was similar to that of the previous part, with the functional tests achieving 76.4% and the combined set achieving 99.3%.

The I_{DDQ} tests were separately generated and applied using the scan chain. No measurements were made while the functional vectors were applied and the current limit set was 30 μA under nominal voltage conditions. It would have been desirable to set the current limit less than this value, but the presence of a pullup reduced the available resolution. The measurements were carried out after parts were subjected to voltage stressing (running the functional vectors with V_{DD} at 7 V). The method of measurement was to use the PMU of the tester. In general, this method is excessively time consuming and there are better methods [23] available, but for experimental purposes the increased test time was acceptable. To obtain a distribution of currents over all parts, a separate current measurement was made for a single vector, independently of the I_{DDQ} tests. The production program was modified so that each type of test was presented to the part even if a failure occurred in some other portion of the test.

4.2. Results

A total of three wafer lots was analyzed, consisting of 26,415 die (excluding parts which fail initial continuity and parametric tests). The distribution of static currents

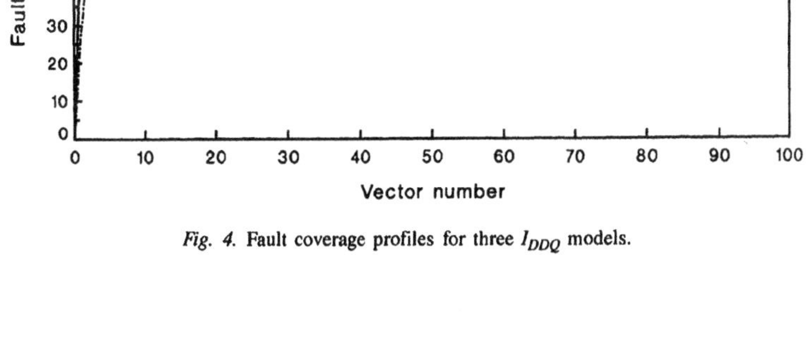

Fig. 4. Fault coverage profiles for three I_{DDQ} models.

(measured using a single vector) for these parts is shown in Figure 5, where each lot is shown with a different bar. Although the majority had currents in the 100–200 nA range, many had significantly higher values than this, ranging up to tens of mA. The second lot had a substantially larger number of parts with high currents. A higher proportion of this lot failed other tests also, which will be discussed later.

A number of comparisons are considered with respect to the relative number of die passing or failing each type of test. For comparison we take the residual reject rate to be zero. The true value would be added to the numbers derived here. For the majority of the comparisons we grouped all three lots together.

The tests were divided into two classes: voltage and current. The voltage tests were further subdivided into functional and scan. The data could then be binned into one of 8 groups, depending on which of the tests it passed or failed. Table 5 shows the number of die in each of the 8 groups. For example, 122 die passed the functional test, but failed both scan and I_{DDQ}. This information is also shown in Venn diagram form in Figure 6. This diagram is not to scale due to the wide variation in sizes of each category, but it visually depicts the distribution of chips.

Table 6 is similar in structure to Table 5 and shows the reject rates for various combinations of the tests. If no testing at all were performed, the reject rate would be 16.46%, while if just functional and I_{DDQ} were performed, it would be 0.09% (again, ignoring the residual reject rate). Table 6 shows that the most effective test of the three is the I_{DDQ} test, which by itself gives a reject rate of only 0.8%. The next most effective is the scan test, with a reject rate of 6.0%, and the least effective is the functional test, with a reject rate of 6.4%.

4.2.1. Stuck-at versus I_{DDQ}. We distinguish here the mutual exclusivity of die rejected by stuck-at and/or I_{DDQ} tests. In the general sense, the intention is to compare voltage sensing with current sensing. In this regard both the scan and functional tests represent voltage testing. However, previous work has limited simulation and test chip data to include only automatically

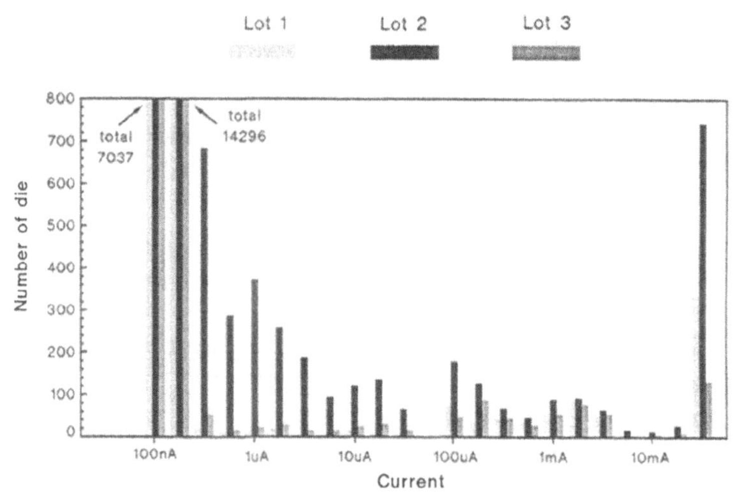

Fig. 5. Current distribution for tested die based upon a single vector.

Table 5. Distribution of die by test.

Number of Die	Scan and Functional Tests				
	Pass-Pass	Pass-Fail	Fail-Pass	Fail-Fail	Total
Pass I_{DDQ}	22066	25	19	134	22244
Fail I_{DDQ}	1358	36	122	2655	4171
Total	23424	61	141	2789	26415

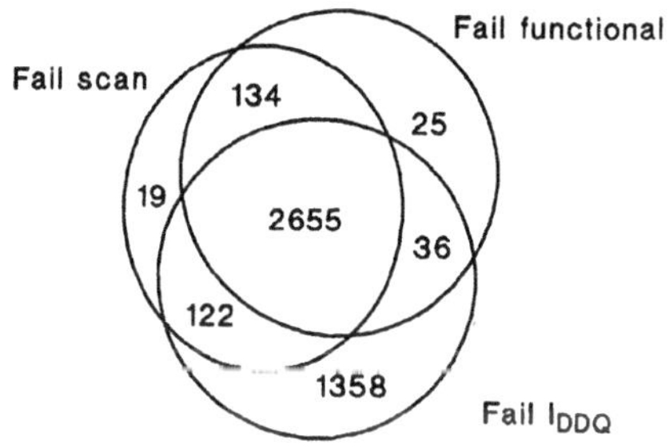

Fig. 6. Distribution of failing die in each test class.

Table 6. Reject rate for various tests.

	Scan and Functional Tests			
Reject Rates (%)	Neither	Noscan/ Func	Scan/ Nofunc	Both
Without I_{DDQ}	16.46	6.36	6.04	5.80
With I_{DDQ}	0.80	0.09	0.11	0.00

generated stuck-at vectors. Therefore, to compare our results with other data available [9], [10] we will use the scan vectors as the stuck-at tests.

In our case 64% of defective parts failed both I_{DDQ} and stuck-at. The proportion failing only I_{DDQ} was 32% while those failing only stuck-at was 4%. These results follow the trend of [10] in that a majority fail both tests. Both are at variance with [9], but the latter are for a very simple chip which is predominantly I/O circuitry.

4.2.2. Voltage Failures with High Currents. More general comparisons of voltage and current testing are also possible. Taking scan and functional vectors together, of the parts that failed any of these voltage tests, 94% also exhibited excessive current. The high coverage of the I_{DDQ} tests in the first few vectors resulted in 81% of these I_{DDQ} failures being detected on the first vector.

Given the current distributions in Figure 5, it is not surprising that the second wafer lot yielded less than the other two, with more failures in every category. Interestingly, in terms of percentages of *bad* parts, the distribution of failures in each category of test was very similar for all three lots.

4.2.3. Agreement with Fault Simulation. It is possible to graph the cumulative number of I_{DDQ} fails versus the vector number and compare this with fault simulation. Since we were experimenting with a number of different grading methods, this provides a rough method of evaluating them. The cumulative fails should track the fault coverage profile closely. Figure 7 shows this is indeed the case for the FET shorts/metal bridge coverages. The pseudo-stuck-at coverage rises much more slowly, and the data indicate therefore that this is a poorer indication of the coverage of the types of defects which give rise to elevated I_{DDQ}.

Due to the fast increase in detections by I_{DDQ}, economic arguments might indicate using only a small subset of vectors. For example, 90% of all I_{DDQ} rejects occurred in the first 8 vectors. However, for the purposes of improving quality, the real contribution comes from *unique* I_{DDQ} failures, since nonunique failures can be detected by other tests. Figure 7 also shows the cumulative proportion of unique failures, which rises much more slowly than total failures. Thus, 90% of unique failures are not reached until vector 30. Coincidentally, this curve more closely matches the pseudo-stuck-at coverage curve, but the reason is more likely to be the fact that both represent more difficult faults, rather than that the pseudo-stuck-at model is a more realistic representation of the types of defects that give tise to unique I_{DDQ} failures.

4.2.4. Reject Rate of Functional Vectors. With functional vectors alone, the reject rate (from Table 6) would be 6.4%. However, this reject rate mostly comprises parts which fail only I_{DDQ} tests. Neglecting these, the reject rate drops to 0.6%. Even though they must contain a defect, die which fail I_{DDQ} tests but pass all others may function correctly if installed in their destination boards and/or systems. It is believed, however, that these parts may be a reliability risk [4], [6]. Evidence of potential problems is indicated in Figure 8, which shows the current distribution of parts which uniquely fail I_{DDQ}. It is disturbing to observe parts with 10's of mA of current which pass every other test, since without any I_{DDQ} testing, these parts will be placed into systems.

4.2.5. Timing Failures and I_{DDQ}. Timing failures occur when a part operates correctly at a slow clock rate but fails when the normal clock rate is applied. A number of different mechanisms (e.g., partially conducting transistors, resistive bridges) have been identified as potentially causing this type of failure, but their manifestation can be subtle and difficult to analyze quantitatively. Efforts have been expended recently in investigating delay faults (some of these are discussed in [24]).

An unanswered question relates to the effectiveness of I_{DDQ} in detecting defects which give rise to delay faults. Some defects which give rise to increased current

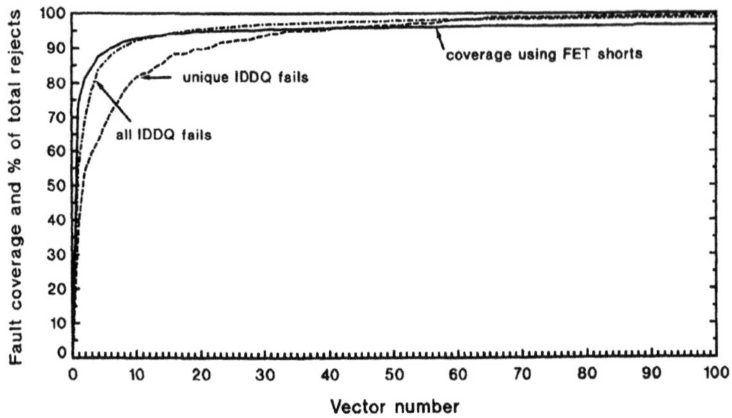

Fig. 7. Cumulative rejected die for both all and unique I_{DDQ} failures.

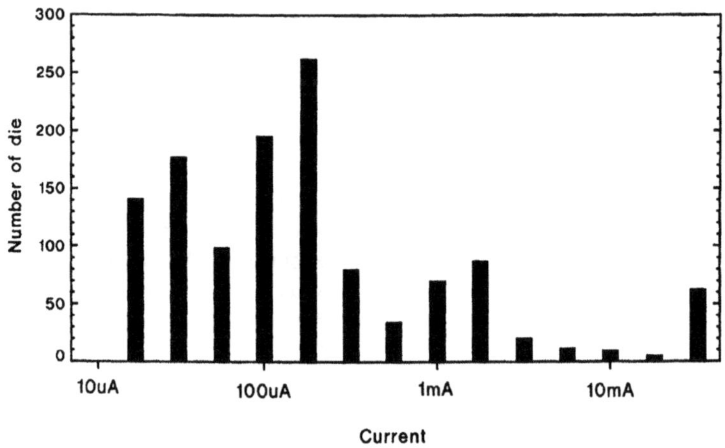

Fig. 8. Current distribution of die which unique fail I_{DDQ}.

could also be seen to result in increased delay. An example is a stuck-on transistor where, because of voltage division, an output signal could be a degraded but correct logical value. This degraded value could alter the propagation delay through the affected gate. Other defects giving rise to this effect are gate oxide shorts, resistive bridges and punchthrough.

The results indicate that there is much less correlation between I_{DDQ} failures and timing failures than between I_{DDQ} failures and voltage failures in general. A total of 21 parts passed all voltage tests except the speed tests. This represents 0.1% of parts that would otherwise be passed as good. Of these 21, 11 passed I_{DDQ} as well. That is, I_{DDQ} testing detected only 10 out of 21 parts which failed only at speed. Since the failing currents on these parts ranged from the threshold up to 10 mA it is possible that the number of detected timing failures would increase if the current limit were decreased, but we were not able to investigate this. Of the 11 parts which passed I_{DDQ}, only 2 had abnormal static currents (1.4 μA and 2.5 μA). Figure 8 shows the I_{DDQ} distribution for parts that uniquely failed I_{DDQ}.

5. Discussion

We have shown that even to maintain a given quality level, fault coverage must increase with increasing chip size (assuming the defect density of the fabrication process remains constant). Since stuck-at tests by themselves are inadequate, the detection of other fault types needs to be quantified in order to obtain a good estimate of total fault coverage.

The additional metrics investigated here were intra-gate transistor shorts and inter-gate bridging. I_{DDQ} testing was very effective in detecting these types of failures. Table 6 indicates that these are still only a subset of the faults detected by the functional vectors. The combination of high stuck-at scan vectors and I_{DDQ} still passes defective parts which are detected by the functional vectors.

Table 6 combines the results for the slow speed and high speed functional vectors. For the first wafer lot the 20 MHz tests were not run at 2 MHz so it is not possible to determine which failures were due to timing and which were due to static faults. However, the two other lots had the additional 2 MHz tests run, so that any failures at 20 MHz or 32 MHz after the die passed the 2 MHz tests, were due solely to timing faults. Since it has been shown that many timing failures were not caught by the I_{DDQ} vectors, an alternative test combination consisting of (scan + I_{DDQ} + timing) was analyzed. A part was characterized as failing the timing portion of this combination only if it passed all the 2 MHz tests. Using this combination, only 7 known defective parts were not detected, which corresponds to 0.046%, or 460 ppm. Of these parts, 2 had static currents in excess of 1.5 μA. If the I_{DDQ} threshold were decreased to this value, the contribution to the reject rate would be no more than 360 ppm. It is quite possible that the other 5 parts also had currents of this magnitude in some state(s), but we were not able to verify this.

The results show that I_{DDQ} tests do a very good but not complete job of detecting defects missed by the scan tests. The analysis of timing failures suggests that a combination of high static stuck-at coverage, I_{DDQ} tests and automatically generated timing tests will result in excellent quality levels, which could be improved still further by targeted bridging tests. This reinforces the need for additional tests which are graded according to different fault types. By adopting such a strategy, reliance on peripheral coverage by functional vectors will be reduced. This could have significant impact as more designs are being produced using synthesis systems. A designer has less knowledge of the structure of a synthesized design than is possible with a design created from schematics. The effectiveness of functional vectors in a synthesis environment is unknown and it is therefore desirable to have a test strategy which does not rely on them.

Acknowledgment

A significant contribution to the work described in this article was made by Vic Johansen and Inshen Chiang, of Hewlett-Packard's California Design Center, who helped plan the experiment and carry out all the production test modifications necessary to facilitate data collection. Many helpful discussions were also held with Dr. C.F. Hawkins, whose input significantly improved this article.

References

1. F.J. Ferguson and J.P. Shen, "Extraction and simulation of realistic CMOS faults using inductive fault analysis," *Proc. Int. Test Conf.*, pp. 475–484, September 1988.
2. P.C. Maxwell, R.C. Aitken, V. Johansen, and I. Chiang, "The effect of different test sets on quality level prediction: When is 80% better than 90%?" *Proc. Int. Test Conf.*, pp. 358-364, October 1991.
3. C.F. Hawkins, J.M. Soden, R.R. Fritzemeier and L.K. Horning, "Quiescent power supply current measurement for CMOS IC defect detection," *IEEE Trans. on Industrial Electronics*, vol. 36, pp. 211-218, May 1989.
4. C.F. Hawkins and J.M. Soden, "Reliability and electrical properties of gate oxide shorts in CMOS ICs," *Proc. Int. Test Conf.*, pp. 443-451, September 1986.
5. J.M. Soden, R.K. Treece, M.R. Taylor and C.F. Hawkins," "CMOS IC stuck-open fault electrical effects and design considerations," *Proc. Int. Test Conf.*, pp. 423–430, August 1989.
6. S. McEuen, "IDDq benefits," *VLSI Test Symposium*, pp. 285–290, April 1991.
7. F.J. Ferguson, M. Taylor and T. Larrabee, "Testing for parameteric faults in static CMOS circuits," *Proc. Int. Test Conf.*, pp. 436–443, September 1990.
8. R.R. Fritzemeier, J.M. Soden, R.K. Treece and C.F. Hawkins, "Increased CMOS IC stuck-at fault coverage with reduced I_{DDQ} test sets," *Proc. Int. Test Conf.*, pp. 427-433, September 1990.
9. T. Storey, W. Maly, J. Andrews and M. Miske, "Comparing stuck fault and current testing via CMOS chip test," *Proc. European Test Conf.*, pp. 149-156, April 1991.
10. T.Storey, W. Maly, J. Andrews and M. Miske, "Stuck fault and current test comparison using CMOS chip test," *Proc. Int. Test Conf.*, pp. 311-318, October 1991.
11. R.H. Williams, R.G. Wagner and C.F. Hawkins, "Testing errors: data and calculations in an IC manufacturing process," *Proc. Int. Test Conf.*, pp. 352-361, September 1992.
12. J.A. Cunningham, "The use and evaluation of yield models in integrated circuit manufacturing," *IEEE Trans. Semiconductor Manuf.*, vol. 3, pp. 60-71, May 1990.
13. B.T. Murphy, "Cost-size optima of monolithic integrated circuits," *Proc. IEEE*, pp. 1537-1545, December 1964.

14. C.H. Stapper, "Defect density distribution for LSI yield calculations," *IEEE Trans. Electron Devices*, vol. ED-20, pp. 655–657, July 1973.
15. C.H. Stapper, "Large-area fault clusters and fault tolerance in VLSI circuits: A review," *IBM J. Res. Develop.*, vol. 33, pp. 162–173, March 1989.
16. C.L. Henderson, J.M. Soden and C.F. Hawkins, "The behavior and testing implications of CMOS IC logic gate open circuits," *Proc. Int. Test Conf.*, pp. 302-309, October 1991.
17. W. Maly, P.K. Nag and P. Nigh, "Testing oriented analysis of CMOS ICs with opens," *Proc. Int. Conf. on CAD*, pp. 344–347, November 1988.
18. Y.K. Malaiya and S.Y.H. Su, "A new fault model and testing technique for CMOS devices," *Proc. Int. Test Conf.*, pp. 25–34, November 1982.
19. W. Mao, R.K. Gulati, D.K. Goel, and M.D. Ciletti, "QUIETEST: A quiescent current testing methodology for detecting leakage faults," *Proc. Int. Conf. on CAD*, pp. 280–283, November 1990.
20. R.C. Aitken, "Fault location with current monitoring," *Proc. Int. Test Conf.*, pp. 623–632, October 1991.
21. A. Jee, "Carafe: An inductive fault analysis tool for CMOS VLSI circuits," Technical Report UCSC-CRL-91-24, University of California at Santa Cruz, Computer Engineering Dept., 1990.
22. R. Rodriguez-Montanes, J.A. Segura, V.H. Champac, J. Figueras and J.A. Rubio, "Current vs. logic testing of gate oxide short, floating gate and bridging failures in CMOS," *Proc. Int. Test Conf.*, pp. 510–519, October 1991.
23. M. Keating and D. Meyer, "A new approach to dynamic IDD testing," *Proc. Int. Test Conf.*, pp. 316–321, September 1987.
24. S. Devadas and K. Keutzer, "Synthesis of robust delay-fault testable circuits: Theory," *IEEE Trans. CAD*, vol. 11, pp. 87–101, January 1992.

Peter C. Maxwell is with the Design Technology Center of Hewlett-Packard's Integrated Circuits Business Division. He received the B.Sc. and M.Sc. (Honours) degrees from the University of Auckland, New Zealand in 1966 and 1968, respectively, and the Ph.D. degree from the Australian National University, Canberra, Australia in 1972. From 1972 to 1988 he was on the faculty of the University of New South Wales, Sydney, Australia, prior to moving to California. During this period he spent time as a Visiting Professor at the University of California, Berkeley and San Diego.

His research interests include test methodologies, design for testability and fault modelling for both test generation and diagnosis purposes. Dr. Maxwell is a Senior Member of the IEEE.

Robert C. Aitken received the B.Sc. (Honours) and M.Sc. degrees in computer science from the University of Victoria, Canada in 1985 and 1986, respectively, and the Ph.D. degree in electrical engineering from McGill University, Montreal, Canada in 1990.

Since 1990 he has been a member of the technical staff at Hewlett-Packard's Design Technology Center in Santa Clara, California. His research interests include fault diagnosis, fault modelling, design for testability, and built-in self-test. Dr. Aitken is a member of the IEEE.

JOURNAL OF ELECTRONIC TESTING: Theory and Applications, 3, 317-325 (1992)

I_{DDQ} Testing in CMOS Digital ASICs

ROGER PERRY
Storage Technology Corporation, 2270 South 88th Street, Louisville, CO 80028-5110

Received February 12, 1992. Revised August 13, 1992.
Editor: C.F. Hawkins

Abstract. I_{DDQ} testing with precision measurement unit (PMU) was used to eliminate early life failures caused by CMOS digital ASICs in our products. Failure analysis of the rejected parts found that bridging faults caused by particles were not detected in incoming tests created by automatic test generation (ATG) for stuck-at-faults (SAF). The nominal 99.6% SAF test coverage required to release a design for production was not enough! This article shows how I_{DDQ} testing and supplier process improvements affected our early life failure rates over a three year period. A typical I_{DDQ} measurement distribution, effects of multiple I_{DDQ} testing, and examples of the defects found are presented. The effects of less than 99.6% fault coverage after the I_{DDQ} testing was implemented are reviewed. The methods used to establish I_{DDQ} test limits and implement the I_{DDQ} test with existing ATG testing are included. This article is a revision of one given at International Test Conference [1].

Keywords: Automatic test program generation, bridging fault, early life failures, I_{DDQ} current, stuck-at fault.

1. Introduction

During the first half of 1988 we started production of a major program and worked to ensure the highest quality levels for our CMOS ASIC designs. A stringent design release requirement of 99.6% SAF coverage was established. We 100% tested all incoming ASICs and did extensive reliability testing. By third quarter of 1988 we knew our test strategy had failed. We were barely achieving 10,000 ppm failure rates against a goal of 100 ppm. The system failures continued to increase and reached an unacceptable level by fourth quarter 1988.

2. I_{DDQ} Test Method

Failure analysis at the ASIC supplier showed that all rejects had one thing in common, higher than normal I_{DDQ} leakage due to particles in the inter-metal dielectric. We introduced measurement of the quiescent V_{DD} supply current, I_{DDQ}, to our incoming ASIC testing to help reduce early life failures in our products. The affect of I_{DDQ} testing on our product line follows.

2.1. CMOS Design and Automatic Test Generation

Our CMOS digital ASIC designs use scan design techniques to achieve a nominal 99.6% stuck-at-fault (SAF) coverage. The designs use custom level and edge sensitive scan latch macros. The scan latches use a master and slave clock to gate signals from marco to macro in an internal scan chain(s) to ensure access to all nodes. The two nonoverlapping clocks eliminate race conditions. The scan path circuitry is disabled during normal device function. A fault simulator runs until all input and output faults are graded or the backtrack limit is reached. The designer reviews all undetected nodes and redundant paths and makes changes to the design scan path(s) as needed to increase the fault grade. The ASIC specification targeted a SAF coverage of 99.6%. The 13 designs presented here ranged in size from 2.5k to 5.6k gates with the average being 4k gates. All of the designs are in pin grid array (PGA) packages with 84 and 120 pin counts.

When the design is complete, an automatic test pattern generator (ATG) creates a test program for a specific test system. The ATG programs run at a 1 MHz rate and range in size from 50k to 200k vectors. The scan-in operation is completed with a parallel vector where all inputs are set and produce known outputs on all pins. The next scan operation combines the scan-out for the first test with the scan-in for the next test. The sequence repeats with the last test being a scan-out operation. During the scan operation only the scan-in and scan-out pins are functional. The other inputs are held fixed and the nonscan outputs are in a don't care state.

2.2. *Incoming Test*

An incoming test was setup in 1987 because our test programs required extended local memory (ELM) that was not available in the production testers at the ASIC supplier. We 100% tested all incoming parts at a third party test house using Sentry 21 testers with 256K of ELM. Throughout most of 1988 the incoming test for I_{DDQ} had a 10 milliampere (mA) limit with a ± 2 mA accuracy and had no history of rejects.

2.3. *I_{DDQ} Test Revision*

The line fallout increased steadily during 1988 and reached an unacceptable level by the 4th quarter of 1988 (Q4–88). The common symptom of the failing parts was higher than normal I_{DDQ} current. The existing 10 mA I_{DDQ} test was too gross to detect the sub 1 mA currents measured for *good* parts and the limit was set too high to detect the 2–8 mA I_{DDQ} currents measured in failure analysis for *bad* parts. Our first I_{DDQ} test was implemented in three weeks using a simple test method. We modified the test hardware and put in a relay so we could power up the part to 5.5 V using a PMU hooked to a spare tester channel. Test vectors were chosen by cycling the tester through several hundred gross functional test vectors, stopping, and measuring the I_{DDQ} current. The I_{DDQ} vector test points for the designs ranged from functional vector number 293 to 1160 with the average at 775. This put the I_{DDQ} test point after the first complete internal scan in operation was complete. If the current was below 1 mA, we had a test point. Otherwise, we tried more points in the functional test and picked the one that had the lowest current. We reduced the input leakage current by setting the input voltage low, *VIL*, to ground (V_{SS}) and input voltage high, *VIH*, to supply voltage (V_{DD}) during the test.

2.4. *I_{DDQ} Test Limit*

The I_{DDQ} test limits were determined by data logging several hundred parts per design, one I_{DDQ} measurement per part. Measurements of I_{DDQ} for good CMOS Digital ASICs were found to be in the 40 to 3000 microampere (μA) range depending on the design. The differences in nominal I_{DDQ} currents were due to input leakage currents through pull-up resistors and output currents through load resistors on the test fixture. Figure 1 shows the combined I_{DDQ} measurements for three designs. The distribution is based on I_{DDQ} measurements for 1057 parts and shows the characteristic long tail of I_{DDQ} currents we found for all of the designs. The I_{DDQ} measurements range from 310 μA to 1740 μA. The first two bins with currents ranging from 300 to 500 μA contain 94% of the parts. The remaining 6% of the parts ranged from 500 μA to 1.8 mA.

Our first attempt at an I_{DDQ} upper control test limit, UCL, was calculated using $UCL = X + 3 * S$ where X is the mean and S the standard deviation. This limit was abandoned when it became obvious that the limit was grossly inflated by the magnitude of currents found in the distribution tail. Instead, we used a limit based on the distance between the 25% and 75% points on the current distribution, the interquartile range statistic $R_{1\text{-}3}$. The calculated upper limit for I_{DDQ} was $UCL = M + 3 * R_{1\text{-}3}$, where M is the median. This method results in *reasonable* I_{DDQ} test limits that are not inflated by the magnitude of the current measured in the tail of the distribution. The difference between the two limits is shown in Figure 1. For this distribution the interquartile limit is, 550 μA, with 5% of the parts failing the I_{DDQ} limit. The standard deviation limit is 960 μA with 2.5% of the parts being rejected. The 550 μA limit was applicable only to these three designs.

The new I_{DDQ} limit of 550 μA was fixed in the incoming test programs and we went back to *pass/fail* I_{DDQ} testing. The I_{DDQ} test was at room temperature with V_{DD} at 5.5 V. The measurement repeatability was ±20 μA. The average reject rate for the new I_{DDQ} test limit was 2.0%. Failure analysis of the test rejects found the I_{DDQ} currents were due to the same particles in the inter-metal dielectric found in our system failures. Limits were adjusted periodically to ensure the I_{DDQ} rejects were not due to a shift in the ASIC suppliers process.

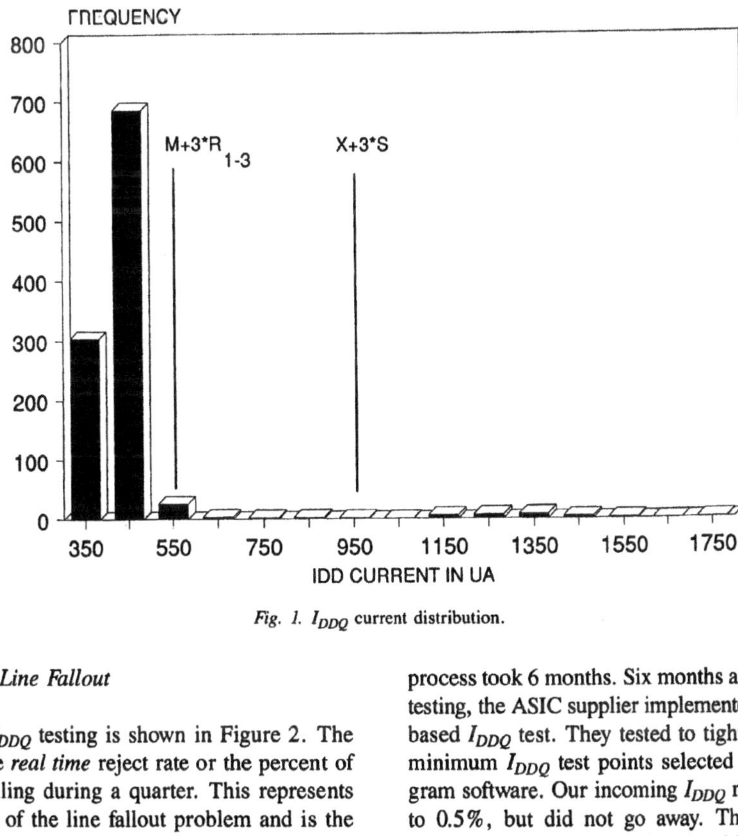

Fig. 1. I_{DDQ} current distribution.

2.5. *Effect on Line Fallout*

The effect of I_{DDQ} testing is shown in Figure 2. The solid line is the *real time* reject rate or the percent of total rejects failing during a quarter. This represents our perception of the line fallout problem and is the sum of all rejects throughout our system test process. The failure rate was peaking at the time we implemented our testing and dropped after introducing the I_{DDQ} testing.

The long term effects of the I_{DDQ} testing and process improvement are better presented by plotting the percent of total rejects against the manufacturing date code and the SAF coverage (the solid and cross hatched bars, Figure 2). The line reject rate by manufacturing date code indicates a stable defect rate for 1988 and 1989. Four months after we started the first I_{DDQ} test, the inter-metallic dielectric process was changed to eliminate the source of inter-metal particles. The transition to where all incoming material was from the new process took 6 months. Six months after we started our testing, the ASIC supplier implemented their own PMU based I_{DDQ} test. They tested to tighter I_{DDQ} limits, at minimum I_{DDQ} test points selected by their test program software. Our incoming I_{DDQ} reject rate dropped to 0.5%, but did not go away. The line reject rate dropped to an acceptably low level by Q1-90 and has remained constant.

2.6. *System Test Procedures*

The line fallout is measured from system level testing through to reliability returns. The initial card test, where assembly related defects are repaired is not included in the total. The test process starts with a box level test where a completed subassembly is powered up and cycled through diagnostic test. This is the first point where the ASICs are tested at speed (nominal 4 MHz). Most of the rejects presented here come from

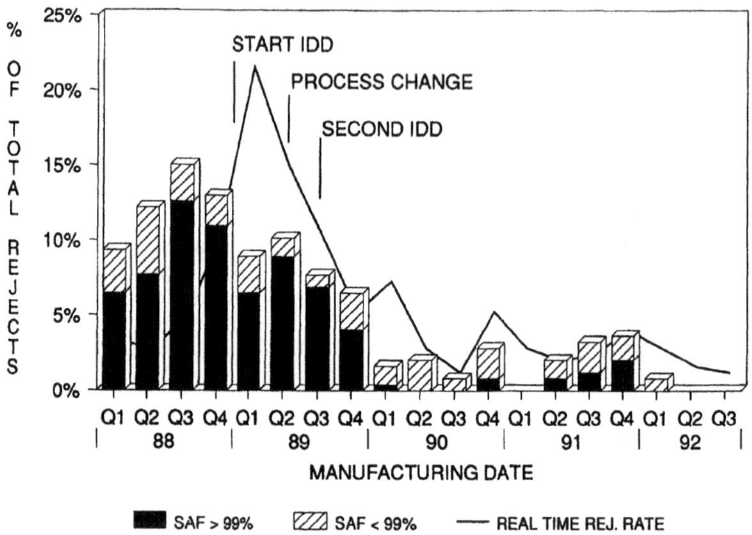

Fig. 2. Effect of I_{DDQ} testing on line fallout.

the box test level. The next test level is where the subsystems are assembled into a functional system. This system level of testing includes data processing, system mechanical checks and diagnostics. The system is tested continuously for several days. The system is then taken through an elevated temperature and mechanical vibration. The system level test is repeated to locate any induced failures. After the second system level test, the passing systems are moved into finished goods for shipment.

All cards failing the box and system tests were validated in a second system in the failure analysis lab before a component is removed. All ASIC failures were submitted for internal component failure analysis with field failures and final system test failures having top priority. Depending on prior history for a manufacturing date code, samples of the failures were submitted for failure analysis to the ASIC supplier. Since 1990, all failures have been submitted for failure analysis.

In summary, the majority of rejects in Figure 2 are functional failures with very few misdiagnosed rejects. The line reject data are a good tool showing how well a group of components work, but it is not a controlled experiment and is not 100% accurate. It should also be noted that the percent of total reject data is affected by the total number of parts used. The increase in rejects in Q3-91 was due in part to a manufacturing ramp up just prior to switching to surface mount manufacturing in 1992. Some of the failing parts from this period are in failure analysis at this time.

2.7. *Effects of Less than 99% SAF Test Coverage*

The cross hatched area in Figure 2 refers to designs with less than 98.5% SAF coverage. Most of the line fallout in the period Q1-90 through Q4-92 came from three designs. When these parts were submitted for failure analysis, no defects were found. They passed all ATG and I_{DDQ} electrical tests. During 1990 the ATG program was revised to change the fault grading methods. The primary change was to include unobservable faults in the tristate control and slave clocks in the total fault count. The design release rules were also changed so input faults cannot be signed off even if the design had reached 99.6% SAF. These changes resulted in a drop in the SAF percentage coverage for some of designs after they were released. We were able to increase test coverage on one design from 95% SAF to 99.6% by redoing the ATG test program.

Table 1 shows the designs by SAF, percent of total rejects, and percent reject rate calculated from the number of rejects divided by the total number of parts used. All of the data are for the line fallout 1990 to the present (5/92).

Figure 3 shows a scatter plot of percent rejected versus the delta SAF, (100%-SAF) for each design. The trend line in Figure 3 shows the general effect of SAF coverage on percent of total line fallout. The percent reject is calculated by taking the number of line rejects divided by the total number of parts used. The high percent reject rate for the CP design is due to a limited

Table 1. Stuck-at-fault % by design.

Design	SAF % new (old)	% Total Reject	% Reject
DW	91.0 (98.1)	6.3	0.058
CR	98.0 (99.6)	2.0	0.118
CI	98.0	2.0	0.054
CK	98.75	0.8	0.053
CB	99.16	1.2	0.073
CG	99.26	0.8	0.028
CL	99.38	0.4	0.027
CA	99.44	0.0	0.000
CP	99.47	0.8	0.101
CT	99.52	0.4	0.023
CE	99.52 (95.0)	0.0	0.000
CJ	99.60	0.0	0.000
CF	99.65	0.4	0.028

number of parts being used. The 91% SAF data point was not included in the trend line. The percent reject data in Figure 3 indicate it behaves more like its original SAF than the revised SAF. The high percentage of total line fallout for this design is due to the large number of parts used. The fault score design was lowered due to a large number of slave clock undetectable faults. The total percent rejection for all designs is 0.043%. Control chart analysis showed 5 of 7 designs with SAF less than 99.4% were over or near at the upper control limt (U Chart with a multiplier of 1). The remaining designs have had 1 of 6 over the limit. Three of the designs have had no line fallout of a population of 19,000 parts. [2] has a detailed review of rejects versus SAF test coverage that provides a better method for determining the test coverage versus SAF.

2.8. *Summary Effect I_{DDQ} Testing on Line Fallout*

We have not had a line reject for the last two years for this product line (as of 5/92) caused by a point defect from the ASIC supplier. Our first attempt at implementing I_{DDQ} tests was crude, but effective. Equal credit for the improvement in the line fallout rate should be given to the ASIC supplier for having successfully made process improvements that eliminated the primary source of the bridging particles.

3. Review of Bridging Faults

We have used the last three years to refine our I_{DDQ} test methods and implement multiple vector I_{DDQ} testing at two ASIC suppliers. During this time we have answered many questions raised by the original I_{DDQ} testing.

3.1. *Bridging Faults*

The first question that came up was why is 99.6% stuck-at-fault coverage not good enough? The answer is

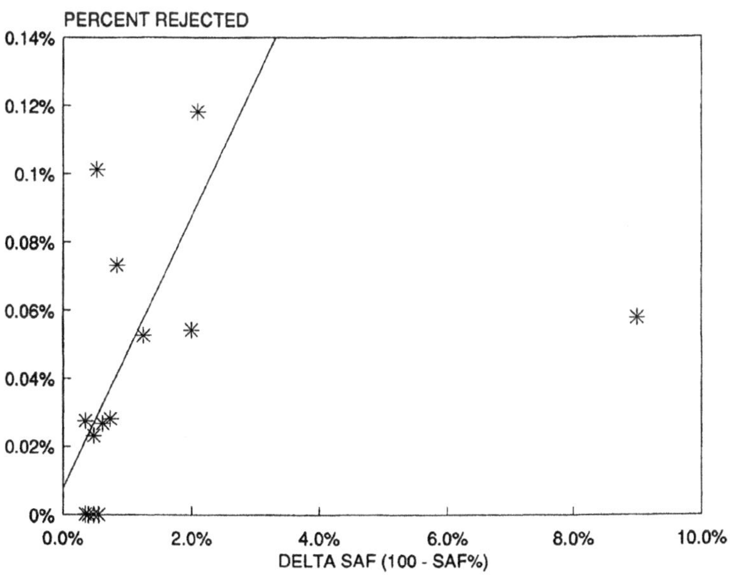

Fig. 3. SAF coverage vs. % line fallout.

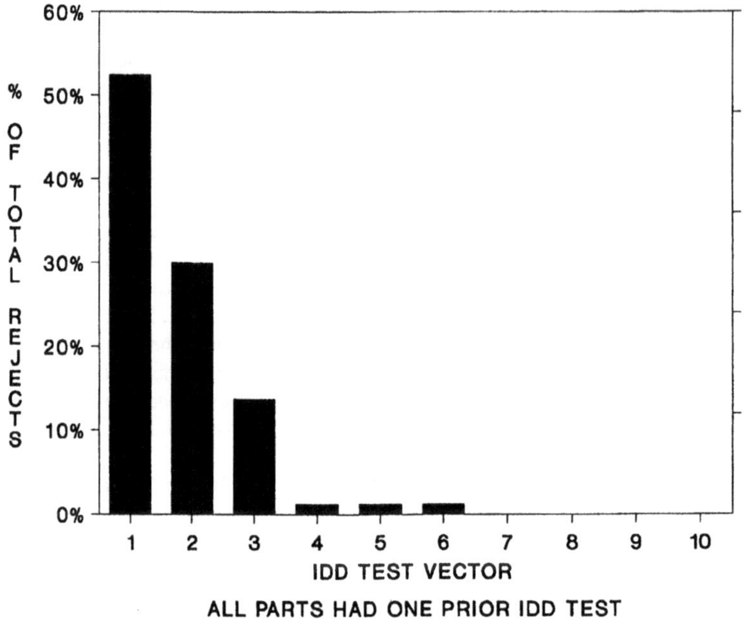

Fig. 4. Multiple I_{DDQ} testing.

bridging faults are not 100% detected by SAF testing. This question has been answered successfully in a number of earlier test articles [2]–[6]. In our two-pass I_{DDQ} testing, we found a total I_{DDQ} reject rate of 2.5% which is in good agreement with the 2%–5% bridging faults not detectable with SAF coverage noted in Table 2 of [3].

3.2. *Multiple I_{DDQ} Test Results*

The second question that came up is how many I_{DDQ} tests are necessary? I_{DDQ} testing is a parallel test. If a bridging defect causes an increase in the I_{DDQ} current, it is detected without having to propagate the fault to an output. In approximate terms, you get 50% of all possible bridging combinations the first time you do an I_{DDQ} test. The problem is in detecting the remaining 50% of possible faults with the minimum number of test vectors. Fortunately, as Figure 4 indicates, the roll off in I_{DDQ} rejects after the first I_{DDQ} test is very sharp. These reject data were obtained from testing a total of about 10,000 surface mount parts from 6 designs. We redesigned 6 of the designs in Table 1 to be put into 84 pin PLCC (plastic lead chip carrier) packages. The gate counts and scan requirements were changed to increase the SAF to 99.6% or better. Testing was done at room temperature on a HP82000 in our failure analysis lab. Testing was stopped after we computed implementation of a multiple vector I_{DDQ} test at the ASIC supplier.

We were not looking at an untested population in Figure 4. All of the parts had received an I_{DDQ} test with a 80 μA guard band over the nominal I_{DDQ}. Our test used a 25 μA guard band so a portion of the total population is evident. Twenty percent of the rejected parts could be detected on power up and showed no variation in I_{DDQ} current at the different test points. The remaining parts had a defect that depended on internal logic levels or toggled inputs and required multiple vector I_{DDQ} tests to be detected. The average number of rejects detected by each I_{DDQ} test vector was 54.5% $\pm$ 6.2% of the total defects found. We projected that we will achieve 50 ppm or better early life failure rates on our new products using multiple vector I_{DDQ} testing without resorting to burn-in or incoming testing.

3.3. *Bridging Test versus SAF Test*

The comparisons of I_{DDQ} bridging fault test versus SAF test presented in [4] and [5] point out substantial advantages in reducing the total number of test vectors by using I_{DDQ} bridging fault vectors. Both articles point out that you can also get defects not detectable by normal SAF methods. It has been our experience that it takes less tester time to complete a 256K vector gross functional test than it takes to do one I_{DDQ} test on the PMU. We run the ATG stuck-at-fault tests first to reduce the number of I_{DDQ} rejects and reduce the total amount of tester time. Both tests are essential for obtaining a low line fallout rate. By integating our I_{DDQ} with the ATG scan testing we feel that we have achieved the quality levels essential to our products.

3.4. Failure Analysis of I_{DDQ} Test Rejects

On-going failure analysis of I_{DDQ} test rejects in our lab and at our ASIC suppliers have identified a broad spectrum of processing faults. We have seen bridging metal and polysilicon, damaged metal, electrical over stress (from a tester), inter-metal shorts from particles, oxide shorts, and gate level defects in the silicon. In one design we noted several floating nodes that affected the I_{DDQ} current by causing longer delay times to reach the current.

Figure 5 shows a typical *bridging* particle causing a metal bridge. This defect was found with I_{DDQ} testing and was easily detected in failure analysis by liquid crystal techniques.

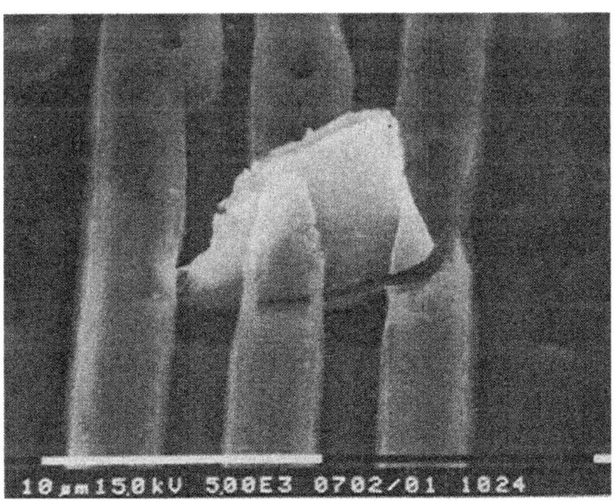

Fig. 5. Bridging particle in metalization.

Figure 6 shows a polysilicon bridge detected with I_{DDQ} testing that affected design nodes at A, D, and E in the photo. The photographs are from Steve Anderson in our failure analysis lab. One benefit of I_{DDQ} testing is that it is possible to quickly locate many of the defects in failure analysis and give direct feedback to the manufacturing lines for corrective action.

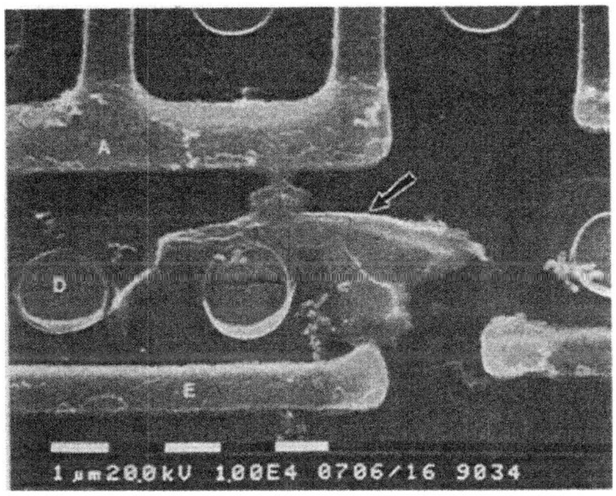

Fig. 6. Bridging polysilicon.

4. I_{DDQ} Test Implementation

We have implemented multiple vector I_{DDQ} tests at both our ASIC suppliers for all our surface mount parts. We have also implemented the I_{DDQ} tests in our failure analysis lab and are in the process of correlating our measurements on rejects. The basic test hardware remains the same. We still use a PMU to make a precise I_{DDQ} measurement. The only hardware change made was to switch out any large filter capacitors on the power lines to improve settling time and eliminate ±20 uA of measurement noise.

4.1. Calculated I_{DDQ} Test Limits

The I_{DDQ} test limit is now specified as part of a design release. The core of a good CMOS ASIC is assumed to have no I_{DDQ} current and all of the measured I_{DDQ} current is due to input leakage. (This assumption is valid, if you do not have embedded memory with standby currents or any floating nodes.) The nominal I_{DDQ} current is calculated by totaling the average input leakages for each of the input macro types used in the design. The average input leakages were measured with $VIL = V_{DD}$ and $VIH = V_{DD} - 20$ millivolt (mV). The 20 mV offset avoids powering up the device through an input with a pull-up resistor and eliminates negative I_{DDQ} due to calibration differences between the PMU and the power supplies used for inputs. The I_{DDQ} current for a pulled-up input is 100 to 1000 times higher current than a normal input leakage. Typical leakages for inputs with a pull-up range from 1 μA - 2 μA for *IIH* and depending on the suppliers manufacturing process, *IIL*, can range from an average of 15 μA to 105 μA. It takes only one input with a pull-up to exceed the total input leakage for a 160 pin design without pull-ups.

We negotiated a fixed minimum I_{DDQ} guard band of 50 μA which is added to the calculated nominal I_{DDQ} current for a design to get the I_{DDQ} test limit. the 50 μA guard band was chosen to ensure that defects could be detected with liquid crystal methods in failure analysis and is large enough to avoid rejection of parts at our suppliers due to production tester differences and normal process variation.

4.2. *Error Calculation*

When the nominal I_{DDQ} current is greater than 200 μA it is necessary to increase the guard band to correct for measurement errors. We calculated our measurement error to be 10% of the nominal current. Most of the error comes from process variation in the input pull-ups. The total error was calculated using propagation of errors to combine the process variation, tester measurement error and variation in the input pull-up current due to a ± 5 °C temperature variation during the test.

4.3. *I_{DDQ} Test Vector Selection*

We stop to measure I_{DDQ} at scan boundaries in the ATG test vectors (internal scan chains not external). The scan boundary test points ensure that all internal nodes, inputs, and outputs are in a known state and that the I_{DDQ} test is repeatable. The first 10 scan ring boundaries are used as test points. We tried unsuccessfully to specify test points some 20k vectors into our ATG test and found that locating the same test points at our supplier was a costly and frustrating experience. Test points are not easy to find especially after the vectors had been processed by a test release program and are output in a new format for their tester.

4.4. *Minimum I_{DDQ} Test Vector*

At each test point we stop the tester and impose a test vector that minimizes I_{DDQ}. The minimum I_{DDQ} test vector serves three purposes: it reduces the total I_{DDQ} current by setting inputs with pull-ups high; disables all device clocks; and, tristates all outputs.

The clock pins are disabled to ensure we do not clock the device during the I_{DDQ} test and randomly change the internal state of the device with the minimum I_{DDQ} test vector. Clock control has precedence over setting input pull-ups to a high state.

All of our designs have a test control pin(s) that tristates all outputs and switches bidirectional pins to an input mode. The minimum I_{DDQ} test vector sets the tristate control pin(s). Tristating the outputs eliminates output currents through fixed tester loads. You can achieve the same effect by floating the outputs during the test or switching out any loads if your tester has dynamic loads.

The minimum I_{DDQ} test vector sets all bidirectional inputs to a known state, if they are in an output mode at a test boundary. The remaining inputs are held as they were when the test is stopped. About 10% of the I_{DDQ} rejects were due to faults in the I/O. The less you restrict the inputs with the minimum I_{DDQ} vector the better. The only design rules we have specified to date is that there be no pull-ups on clock inputs or tristate control pin(s) and outside access to set internal RAM blocks to standby mode. We also recommend avoiding pull-ups where possible on the rest of the inputs to ensure that the maximum number of input states can be tested.

Switching to a minimum I_{DDQ} test vector changes the input states and may affect other internal states depending on the design. This means that it is necessary to start at the beginning of the gross functional test for each I_{DDQ} test. A nominal 30 ms delay is used after the minimum I_{DDQ} test vector is set before measurement.

The minimum I_{DDQ} test vector ensures that each I_{DDQ} test can be made with the lowest possible input leakage current against a fixed I_{DDQ} limit. The same designs we were measuring with 350 μA nominal I_{DDQ} currents in our original tests are now being measured against 75 μA limits with 1 μA accuracy.

5. Conclusion

We have demonstrated with the line fallout data from our system level testing that I_{DDQ} testing, coupled with corrective actions by our suppliers, can substantially reduce early life failures for CMOS digital ASICs. Whether you use brute force to impose a I_{DDQ} test with empirically defined limits, or use a more refined test flow with multiple I_{DDQ} test vectors, results are impressive.

We have been working with our ASIC suppliers, to implement multiple I_{DDQ} testing. Ask for your supplier's support in identifying test points and setting I_{DDQ} test limits for your product. We have found that it is to both parties advantage to do I_{DDQ} testing. We have been able to eliminate incoming testing and they have eliminated a large part of the early life failures without burn-in.

Acknowledgments

The author would like to acknowledge the contributions of our ASIC suppliers. We appreciate the work of their test engineering and failure analysis groups in eliminating our line fallout and incoming test requirements. The author would also like to thank Doug Holmquist, Mike Taylor, Rob Eisenhuth, and Steve Anderson at Storage Technology for their help.

References

1. R. Perry, "I_{DDQ} testing in CMOS digital ASIC's—putting it all together," *Int. Test Conf.*, pp. 151–157, September, 1992.
2. P. Maxwell, R. Aitken, V. Johansen, and I. Chiang, "The effect of different test sets on quality level prediction: When is 80% better than 90%?," *Int. Test Conf.*, pp. 358–364, October 1991.
3. T. Storey and W. Maly, "CMOS bridging fault detection," *Int. Test Conf.*, pp. 842–851, September 1990.
4. R. Fritzemeier, J. Soden, R. Treece, and C. Hawkins, "Increased CMOS IC stuck-at fault coverage with reduced I_{DDQ} test sets," *Int. Test Conf.* pp. 427–433, September 1990.
5. F. Ferguson, M. Taylor, and T. Larrabee, "Testing for parametric faults in static CMOS circuits," *Int. Test Conf.*, pp. 436–443, September 1990.
6. J. Acken, "Testing for bridging faults (shorts) in CMOS circuits," *Des. Auto. Conf.*, pp. 717–718, June 1983.

Roger J. Perry is a Senior Supplier Quality Engineer for Storage Technology Co. His primary responsibility is working with suppliers of digital ASICs and microprocessor microelectronics. Prior to StorageTek, he worked 12 years in the semiconductor industry for GMC Hughes, Burroughs, and National Semiconductor. He has a masters degree in Physics from the University of Arizona.

JOURNAL OF ELECTRONIC TESTING: Theory and Applications, 3, 327-335 (1992)

Reliability Benefits of I_{DDQ}

STEVEN D. McEUEN
Ford Microelectronics, Inc., 9965 Federal Drive, Colorado Springs, CO 80921

Received February 25, 1992. Revised August 13, 1992.
Editor: C.F. Hawkins

Abstract. This is an application-oriented article on I_{DDQ} to guide test engineers to a pragmatic method of implementing I_{DDQ}. This article focuses on different aspects of I_{DDQ} testing; quality, reliability, and test implementation. A description of I_{DDQ} is presented, with different practical methods of implementing it, with empirical reliability data of I_{DDQ} failures, and with empirical burn-in data identifying potential yield benefits. Employing I_{DDQ} testing on digital CMOS technology, the user obtains a product with greater reliability. The data presented within this article, along with increasing customer focus on zero defects, clearly support I_{DDQ} implementation.

Keywords: I_{DDQ}, Gate oxide shorts, reliability.

1. Introduction

The emphasis on quality and reliability of semiconductors saw a dramatic increase in the 1980s. This customer-driven redirection of priorities was, in part, due to pressures by the auto industry, criticism by Hewlett-Packard, and strong off-shore competition [1]. Heading into the 1990s, accepted failure rates range from 1000 dpm to 50 dpm and will decline in the years ahead. In a special edition of Electronic Business, IBM states that by 1994 their goal is to ship products at a failure rate of 3.4 dpm [2]. The terminology used to describe this aggressive approach toward quality is *Zero Defects* or *Six Sigma*. Any company not achieving zero defects in the future will not be competitive in the international or U.S. marketplace.

The problem for semiconductor manufacturers is how to obtain zero defect levels. This is a difficult goal for a standard product supplier and an incredible one for the Application Specific Integrated Circuit (ASIC) and custom supplier. Conventional approaches for accomplishing these goals are: increasing test fault coverage, increasing burn-in coverage, increasing ESD awareness, and/or developing generic qualification vehicles. While these methods are making great strides in quality improvement, I_{DDQ} will be shown to enhance the ability to successfully achieve the desired low defect levels.

Defects or damage can be introduced during the design, layout, fabrication, or assembly of silicon. Defective ICs are always present within the product's population. Substantial progress has and will continue to be made in IC processing and device assembly toward reducing the frequency of occurrence, but devices with defects will still exist. This fact establishes the need for IC testing. Testing identifies faulty components and removes them from the otherwise good population. The most common method of testing ICs cannot adequately discern all the defects that may be present.

The primary model used for identifying test stimulus is the stuck-at fault model. Stuck-at fault modeling's failure to identify all device defects results from the application of a digital test methodology to identify what is basically an analog defect [3]. Defects, when present, are not always severe enough to result in a logical fault. They may be resistive enough to allow the preceding driving source to control the leaking nodes. This is where a testing tool such as I_{DDQ} can help. I_{DDQ} tests the entire circuit by making analog measurements of I_{DDQ}. Testing in this manner allows exposure of devices with defects that normally go undetected.

The ineffectiveness of stuck-at modeling is validated from data publishes by Sandia Labs [3]. 5000 static RAMs were tested of which 433 failed functional testing. Upon testing the same units with an I_{DDQ}

screen, the identical 433 functional failures were identified with an additional 254 (5%) I_{DDQ}-only failures. Additional support as to the ineffectiveness of stuck-at modeling comes from a University of Barcelona study [4]. The conclusion from this study was that I_{DDQ} testing is seven orders of magnitude more sensitive than a logical testing. Failure analysis at Ford Microelectronics Inc. (FMI) has identified field returns that pass 100% stuck-at fault testing, over temperature and voltage, but fail room temperature I_{DDQ} testing [5].

2. Description of I_{DDQ}

I_{DDQ}'s definition is the level of power supply current in a CMOS circuit when all nodes are quiescent (static) [6]. Any abnormal elevation of current indicates the presence of damage or defects. Figure 1 shows how an I_{DDQ} test can identify defects. *Q*4 has a defect that causes a resistive path between its gate and source. When Vin is a logic '1" level, the gate and source of *Q*4 are held at ground by *Q*2, which prevents the I_{DD} current. When Vin switches to a logic "0" level, *Q*4's gate is pulled high by *Q*1 allowing current from V_{DD} through the defective path to ground. The resistance of *Q*4's defect will determine the I_{DDQ} current level and if the circuit is functional.

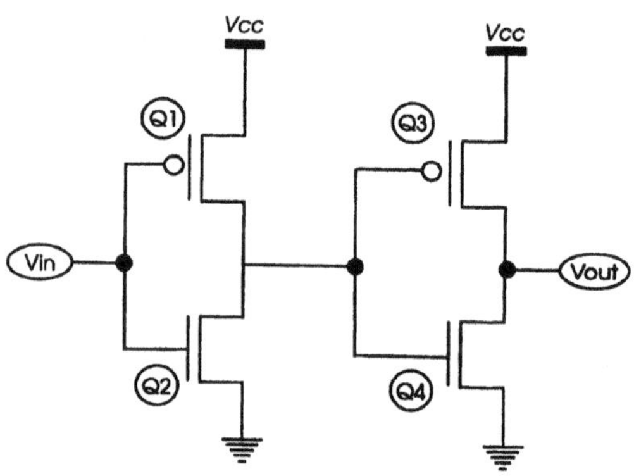

Fig. 1. Example of how I_{DDQ} detects defects.

3. Quality Benefits Derived from I_{DDQ}

The benefits of I_{DDQ} vary depending on implementation, the type of product being tested, and the silicon defect density. If we assume a constant defect density and similar test implementation, then custom products have the most to gain from an I_{DDQ} test. Product volume, product lifetimes, and/or test coverage are generally not adequate on custom material to allow usage of the traditional methods for improving quality levels.

Figure 2 shows that custom devices have the highest percentage of leakage defects that can be identified by an I_{DDQ} test. Removal of leakage failures would eliminate 70% of the field returns. RAMs have the least amount of leakage defects (42%), while EPROMs (44%) and CPU (50.5%) failure levels reside between the two extremes. The failure categories in Figure 2 outline data from device failure analysis reports performed on field returns. All ESD related failures have been removed from this sample and the remaining failures were categorized as:

Leakage = Returns where the failure mechanism results in an unwanted current path between two nodes.

Unknown = Failures that were not identified during failure analysis.

Opens = Failing devices where the failure mechanisms is an open trace.

Others = All other causes for failure.

RAM products have the least to gain from I_{DDQ} testing since RAM testing has empirically evolved over the years to include various test algorithms that are effective in identifying leakage defects. In contrast, most custom device testing developed from functional or stuck-at models that focus on logic functionality and not defect detection.

4. Reliability Benefits Derived from I_{DDQ}

I_{DDQ} testing has been shown to identify gate oxide shorts (GOS) [6]. Many of these gate oxide shorts do not initially cause a functional failure but over time they can deteriorate until functionality is effected. To quantify this effect, FMI conducted a lifetest study on I_{DDQ} failures, recording I_{DDQ} at each read point. It was theorized that any variation in leakage current over time represented an instability that may result in a logic error. Because this instability is unpredictable, the component is considered to have a reliability problem.

Figure 3 illustrates how many of the I_{DDQ} only failures are unstable over time. The population was composed of custom components that had passed all burn-in, hot and cold functional testing, and parametric testing but failed a high temperature I_{DDQ} limit of 10 μA. The population had room temperature I_{DDQ} values ranging from 0 μA to greater than 100 μA. This experiment was repeated and verified on similar devices.

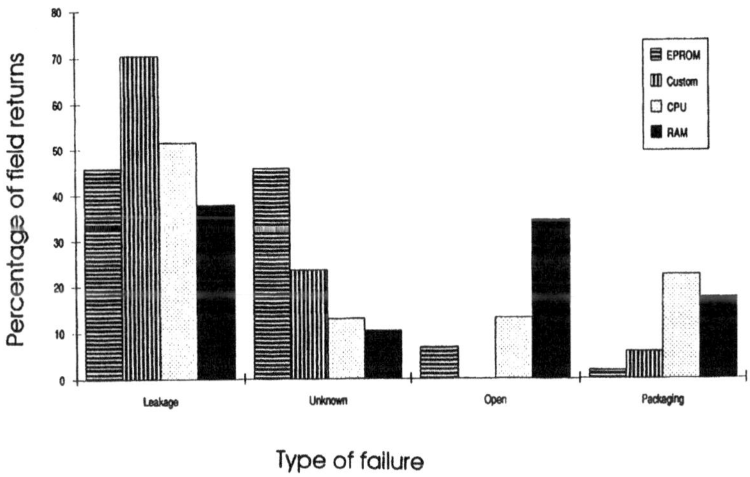

Fig. 2. Histrogram showing I_{DDQ}'s potential effectiveness on identifying field returns.

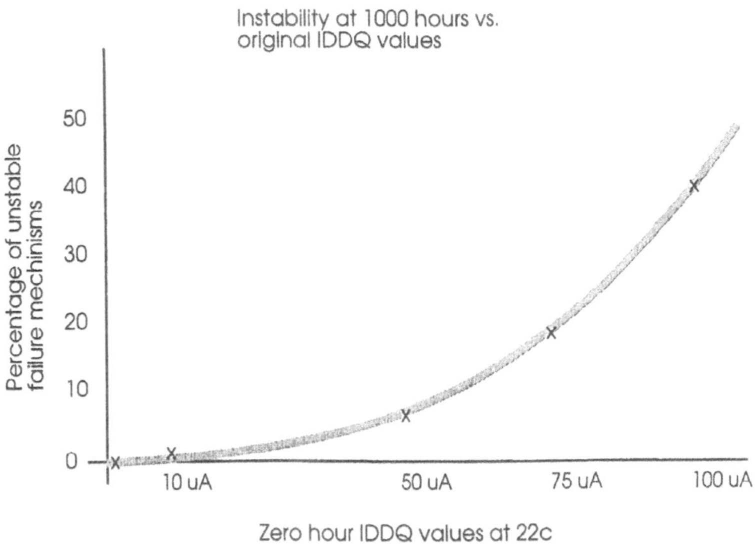

Fig. 3. The percentage of I_{DDQ} failures that exhibit instability after 1000 hours of lifetest vs. the original I_{DDQ} value.

Elevated I_{DDQ} does not necessarily result in nonfunctional behavior. However, data are available confirming that I_{DDQ} failures will result in a reliability problem. Lifetest studies conducted at FMI have shown a statistically significant number of I_{DDQ} failures have become inoperative over time, when compared to their control lots. Figure 4 demonstrates the data from one such study. This lifetest experiment studied the effect of I_{DDQ} values vs. product reliability. Shown is the percentage of I_{DDQ} failures that functionally failed after 1000 lifetest hours. As expected, the most rejects had high initial I_{DDQ} values.

5. Burn-in Benefits Derived from I_{DDQ}

I_{DDQ} has been proven to be beneficial in reducing the number of burn-in failures.[1] A burn-in experiment was conducted on a 7,000 transistor IC using 59,024 die from 26 wafer lots to identify the interdependency

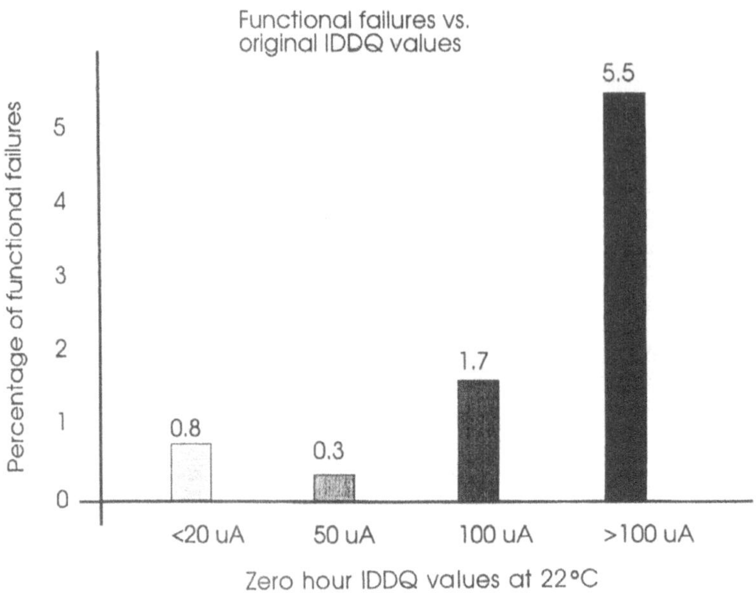

Fig. 4. The percentage of functional failures after 1000 hours of lifetest vs. the original I_{DDQ} value.

between burn-in fallout and I_{DDQ}. The material that was screened out by an I_{DDQ} test prior to burn-in had 51% fewer rejects after burn-in (I_{DDQ} test limit = 10 μA at 22°C). This failure rate reduction was mainly due to a drop in functional failures. This category dropped by two orders of magnitude, while package related failures remained the same.

A separate study was conducted with 6 wafer lots and 21% of the devices that failed only for I_{DDQ} prior to burn-in, failed for functionality after burn-in. A third burn-in comparison study was conducted using a custom 43,000 transistor component, with 7000 units. Material failing only I_{DDQ} was placed into a separate lot prior to burn-in. The overall post burn-in failure rate on the I_{DDQ} failures was 10 times that of the good parts.

Present I_{DDQ} data do not support the elimination of FMI's burn-in. On the contrary, burn-in has aggravated hidden defects that have resulted in new post burn-in I_{DDQ} failures. The primary cause of this is believed to be due to the voltage acceleration present during burn-in [6]–[8]. If the voltage stress seen during burn-in were included in the testing of the device along with an I_{DDQ} screen, then the possibility exists for burn-in to change its form. This new form could then focus more on package related problems such as a thermal cycle test would do.

6. Leakage Mechanisms

There are various causes of leakage. Gate oxide shorts are the most frequently referenced but others do exist. Within this article, leakages are divided into four separate mechanism categories: testing, handling, manufacturing, and design induced.

Testing induced leakage is any leakage that results from the test method. This form of leakage typically is corrected by altering the test conditions. Examples of testing induced leakages are: ATE loading of the DUT's outputs and I/Os, faulty load board, insufficient measurement setting times, or excessive parasitic capacitance.

Manufacturing induced leakage is defined as any defect induced leakage occurring as a consequence of the silicon fabrication or assembly operation. Elimination of this type of leakage is a primary goal of I_{DDQ} because of its inherent reliability implications. Examples of manufacturing induced leakages identified by I_{DDQ} testing are: gate oxide shorts, poly bridges, low transistor thresholds, pn junction leakage, punch through, mobile ionic contamination, stacking faults, particulates, metal bridges, and inner layer oxide shorts. Sandia Labs identified stuck-open faults with an I_{DDQ} test [9].

Handling induced leakages are a result of mishandling. Examples are: ESD and mechanical damage. Both mechanisms are readily detectable with I_{DDQ} after the damage occurs. FMI's experience has been that there are many incoming I_{DDQ} only failures that are identified through failure analysis as having internal ESD or mechanical damage. This damage is not to the extent where logic functionality is affected but the device is considered a walking wounded failure.

In addition, FMI has utilized I_{DDQ} to highlight internal ESD damage during a product's ESD qualification test. I_{DDQ} is an excellent test, since ESD damage usually manifests itself as leakage. External pins have input and output leakage tests that test I/O circuitry adequately. However, I_{DDQ} excels in identifying damage sites that are produced when the energy enters the chip through a power rail and ruptures an internal FET's gate oxide. In some cases, I_{DDQ} testing has been the only indicator that internal ESD damage has occurred.

Design induced leakages are described as any leakage that is present on every device of a particular design. This type of leakage is caused during the design, layout, or mask generation steps and, if not identified, may prevent the implementation of I_{DDQ} testing. Examples are: repeating mask defects, floating gates, incorrect alignment, parasitic devices, internal oscillators, internal resistors, analog circuitry, sense amplifiers, or redundant memory. This form of leakage can be prevented with the use of design guidelines and design rule checks.

7. Implementation of I_{DDQ}

Once the requirement for an I_{DDQ} test has been established, the issues of how to implement it arise. These issues are:

- Test hardware
- Test limits
- Test stimulus
- Test conditions
- Test locations

7.1. I_{DDQ} Test Hardware

The simplest form of hardware is the ATE precision measurement unit. It can be connected to the DUT's power pin and used to measure I_{DDQ}. This strategy is acceptable if the number of measurements is less than 20, otherwise the test time becomes excessive.

A more practical approach is to build an off-chip current monitor such as a series resistor and a FET shunt [10]. The floating V_{DD} node method of Keating and Meyer [11] is more sensitive and can measure sub-μA I_{DDQ} values. The voltage output from these circuits can then be monitored with an ATE high speed comparator. These techniques reduce the time required to make an I_{DDQ} measurement [12]. FMI uses a sample hold approach for measuring I_{DDQ} that allows for adjustments

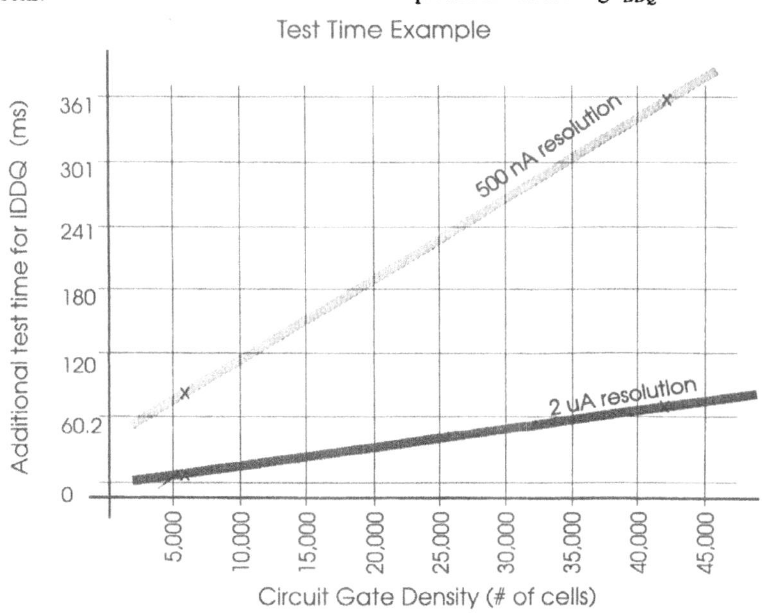

Fig. 5. The lower and upper I_{DDQ} test time limits vs. I_{DDQ} resolution.

in test time based on the resolution required of the measurement circuit. Figure 5 shows two limits depicting the worst and best case $_{DDQ}$ test times. The worst case time of 0.367 s is extremely favorable when compared with the PMU approach that yields a 3.418 second test time for the 40,948 transistor device. Crapuchettes has detailed the design of off-chip I_{DDQ} monitors placed in the tester power supply [12].

One innovative approach uses on-silicon current monitors [13]. This approach, being the most difficult, has the advantages of bypassing leakages caused by the design, reducing the load capacitance, increasing test speed, and with boundary scan, can be used at the system level.

7.2. I_{DDQ} Test Limit

The lack of published data on I_{DDQ} test limits has resulted in a multitude of values. Some values are based on an engineering judgment and others are guesses. A Chauvernet analysis is used at FMI to determine the test limit. This method assumes that any value outside the normal distribution is considered aberrant and rejectable. FMI typically uses a 10 μA to 20 μA I_{DDQ} limit.

7.3. I_{DDQ} Test Stimulus

One I_{DDQ} test method uses 100% nodal coverage [6]. The most common method of achieving this is by testing I_{DDQ} at each test vector. Assuming the IC demands 300,000 functional test vectors to insure 100% fault coverage, it would take 30 s of test time to measure I_{DDQ} at a 10 kHz sample rate.

To reduce test time to an acceptable level, either the number of vectors must be reduced or the current monitor sample rate must be increased. Achieving hardware that can measure low current at high speeds (1 MHz) has not been demonstrated. FMI confronted the test time problem by designing a system that selects a small number of vectors from production test sets to cover gate oxide defects [14]. Test cases have shown that this approach has equivalent leakage fault coverage with less than 1% of the production test vectors as would be achieved by I_{DDQ} measurements on all production test vectors [14]. Using the prior example where the functional vector set was 300,000 vectors and the test time per vector was 100 μs, the use of QUIETEST would reduce the total I_{DDQ} test time from 30 s to 375 ms.

7.4. I_{DDQ} Test Conditions

The amount of current drawn by a defective IC varies with temperature, stress voltage, circuit configuration, and the type of defect. The independent variables are voltage and temperature. A graph of I_{DDQ} vs. V_{DD}, on a failing device, is seen in Figure 6. These data identify a high V_{DD} voltage as the worst case corner for I_{DDQ} testing. Therefore as a minimum, I_{DDQ} should be tested at the highest possible voltage [3], [10].

Figure 7 shows how certain leakages will increase with elevated temperatures. SN1849 has an elevated level and a slight rise in current with temperature. SN1851 starts off at a low current level and has a substantial rise in current. In contrast, parts that do not fail I_{DDQ} at any temperature have a flat current curve. At first glance, high temperature would be considered worst case. However, lifetest investigations at FMI have not found this to be the case. High temperature I_{DDQ} failures, such as SN1851 seen in Figure 7, have not produced significant numbers of functional failures during lifetest. We examined these high temperature I_{DDQ} failures and found that junction leakage was the cause for leakage. Historically FMI has not found that junction leakage worsens with time. Therefore, to keep test effectiveness at a maximum, ambient and/or cold temperature I_{DDQ} testing is recommended. One deviation arises when component specifications require low power battery operation. Under these conditions, high temperature I_{DDQ} testing should be required.

7.5. I_{DDQ} Test Locations

I_{DDQ} has its greatest impact, when placed at probe. This provides immediate feedback to FAB personnel on process defects. Additionally, this prevents the costly assembly of defective material. Besides probe, I_{DDQ} testing is required after burn-in. It is certain that burn-in will activate hidden defects which I_{DDQ} can identify.

7.6. I_{DDQ}'s Future

I_{DDQ}'s future is becoming clearer as additional research data becomes available. Additional benefits are possible once an understanding of the interdependency between I_{DDQ}, functional testing, and burn-in is better understood. In general, the realized and potential benefits of I_{DDQ} testing are:

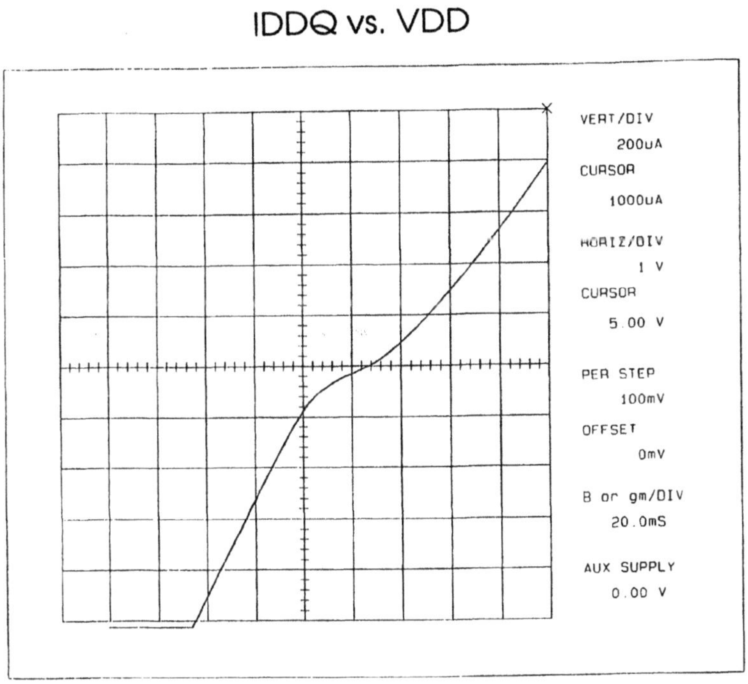

Fig. 6. How I_{DDQ} varies with V_{DD}.

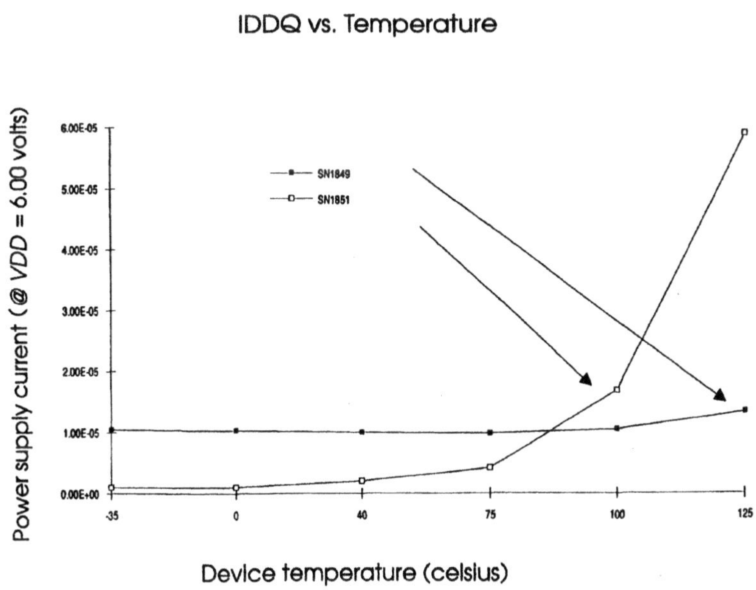

Fig. 7. How I_{DDQ} varies with temperature. SN1851 has a junction leakage problem. SN1849 has an elevated I_{DDQ} value.

- Lower burn-in fallout,
- Replacement of burn-in,
- Usage as a vendor comparison tool,
- Early screening of defects at probe,
- Fewer reliability/customer returns,
- Identifies certain defects that cause delay faults,
- Supplements component AC testing,
- Provides feedback on defects to process personnel,
- Self testing improvements when united with BIST technology,
- Simple customer incoming inspection test,
- Effective screening of ESD walking wounded,
- Identifies bridging faults.

Increasingly, OEMs are requesting information on I_{DDQ} testing. They find that I_{DDQ}, in its simplest form, is easy to implement and it identifies many reliability concerns that would otherwise go undetected. In addition, consumers can use it as a quality comparison tool of multi-source products. With I_{DDQ}, a comparison of leakage defect levels can be made of suppliers without having process information or special test circuits.

The greatest payback for manufacturers is the early screening of defects at probe, and providing feedback to wafer processing personnel, who in most cases have the ability to develop the required corrective actions but are not aware of the problem. This feedback is important for other reasons:

- Most suppliers do not perform failure analysis on normal levels of final test or burn-in fallout. Because of this practice, fabrication engineers do not have access to the data required for root cause corrective actions.
- Fabrication engineers utilize wafer yields as a key indicator for process improvements.
- In general, burn-in and final test failures are similar to the manufacturer's field returns. This similarity can be exploited to provide an early warning indicator to the customers or manufacturer of quality problems.

This author believes that I_{DDQ} can integrate easily with the new test methodologies being implemented on silicon. Utilizing an internal SCAN approach with I_{DDQ} testing can provide a similar leakage fault coverage with less input stimulus when compared with the conventional external pin approach. Utilizing on-silicon I_{DDQ} circuitry with a BIST approach can provide high leakage fault coverage that can be utilized at test or in a field application.

I_{DDQ} has the potential of altering burn-in toward other environmental screens. I_{DDQ} is not foreseen as a replacement for burn-in because of the global effect burn-in has. The temperature acceleration burn-in provides for the aggravation of assembly and packaging related problems is not seen with an I_{DDQ} test. If I_{DDQ} can be proven to identify the die leakage issues, then burn-in could be altered to focus on mechanical and packaging problems.

8. Conclusion

To achieve the high quality and reliability expectations demanded by the marketplace, suppliers need to implement defect oriented screens such as I_{DDQ}. Not taking such action has been shown to place the product at risk. It is recommended, based on the case studies discussed in this article, that I_{DDQ} testing be implemented within the product flow of high quality devices.

Acknowledgments

I would like to thank Murray Lane, Ravi Gulati, Weiwei Mao, Mike Javernick, Deepak Goel, Kevin Healy, Matt Ward, Michelle Miller, and Tom Paquette for their efforts toward the Ford Microelectronics Inc, I_{DDQ} research.

References

1. P. Weiss, "Do we meet world-class quality standards?," *Computer Design/News Edition*, pp. 1–23, September 10, 1990.
2. B. Rayner, "Market driven quality: IBM's six sigma crusade," *Electronic Business*, pp. 68–76, October 15, 1990.
3. L.K. Horning, J.M. Soden, R.R. Fritzemier, and C.F. Hawkins, 'Measurements of quiescent power supply current for CMOS ICs in production testing," *Proc. Int. Test Conf.*, pp. 300–309, September 1987.
4. J.A. Segura, V.H. Champac, R. Rodriguez, J.A. Rubio, and J. Figueras, "On current testing of gate oxide short failure in static memory cell," Internal Report, Department of Engineering, University of Barcelona, Spain.
5. M. Lane and S. McEuen, "I_{DDQ}: A method for detecting potential warranty returns," Ford Microelectronics Inc., Internal Report, August 27, 1990.
6. J. Soden and C. Hawkins, "Test considerations for gate oxide shorts in CMOS ICs," *IEEE Design & Test*, pp. 56–64, August 1986.
7. C. Hawkins and J. Soden, "Reliability and electrical properties of gate oxide shorts in CMOS ICs,'" *Proc. Int. Test Conf.*, pp. 443–451, 1986.
8. J. Soden and C. Hawkins, "Reliability of CMOS ICs with gate oxide shorts," *Semiconductor Int.*, pp. 240–245, May 1987.
9. J. Soden, R. Treece, M. Taylor, and C. Hawkins," CMOS IC stuck-open fault electrical effects and design considerations," *Proc. Int. Test Conf.*, 1989.

10. C. Hawkins, J. Soden, R. Fritzemier, and L. Horning, "Quiescent power supply current measurements for CMOS IC defect detection," *IEEE Trans. on Industrial Electronics*, vol. 36, pp. 211–218, 1989.
11. M. Keating and D. Meyer, "A new approach to dynamic I_{DD} testing," *Proc. Int. Test Conf.*, pp. 316-321, 1987.
12. C. Crapuchettes, "Testing CMOS I_{DD} on large devices," *Proc. Int. Test Conf.*, pp. 310–315, 1987.
13. W. Maly and P. Nigh, "Built-in current testing feasibility study," *Proc. ICCAD*, pp. 340 343, 1989.
14. W. Mao, R. Gulati, D. Goel, and M. Ciletti, "Quietest: A quiescent current testing methodology for detecting leakage faults," *Proc. ICCAD*, pp. 280-283, 1990.

Steven McEuen received his B.S. degree in electrical engineering and computer science from the University of Colorado in Colorado Springs, in 1980. From 1980 to 1985 he worked at INMOS as a Test Engineering supervisor on fast static RAM technology. In 1985 he joined Ford Microelectronics Inc. where he presently is a supervisor in the Reliability Department. He is active in I_{DDQ} research, burn-in systems, power semiconductor failure analysis, power semiconductor technology research, power semiconductor device physics, and stress to failure approaches.

JOURNAL OF ELECTRONIC TESTING: Theory and Applications, 3, 337-348 (1992)

Quiescent Current Analysis and Experimentation of Defective CMOS Circuits

J.A. SEGURA,* V.H. CHAMPAC, R. RODRÍGUEZ-MONTAÑÉS, J. FIGUERAS, J.A. RUBIO
Department of Electronic Engineering, Polytechnical University of Catalonia, Diagonal 649, 08028 Barcelona, Spain

Received Febrary 7, 1992.
Editor: C.F. Hawkins

Abstract. Physical defects widely encountered in today's CMOS processes (bridges, gate oxide short (gas) and floating gates) are modeled taking into account the topology of the defective circuit and the parameters of the technology. These models are used to simulate at electrical level the behavior of a simple 3-inverter chain with a defective inverter. The results are compared with experimental data of integrated circuits fabricated with intentional defects. The influence of the characteristics of each defect on I_{DDQ} has been investigated by electrical simulation and experimentation.

Keywords: Bridging failures, current testing, defect modeling, floating gate opens, gate oxide shorts, intentionally designed defective circuits defects.

1. Introduction

The scope of this work is to characterize the quiescent current behavior of CMOS circuits in the presence of defects in order to evaluate the merits of current testing as a complementary method to voltage (logic) testing.

From the wide spectrum of possible defects reported in CMOS technology we have chosen three defects: one internal to the MOS transistor device, (GOS), and two external to the devices of the circuit (bridges between conducting lines, and opens of the conducting lines of the circuit terminating in a transistor gate). These three defects may be responsible for more than 50% of faulty ICs in CMOS processes, as reported by researchers in the field using Inductive Fault Analysis [1], [2].

Experimentation on the physical behavior of defective circuits is a hard task due to difficulties in identifying circuits representative of the defects investigated. Limited experimental data on quiescent current measures are available at present [3]-[7].

Our objective is investigation of the influence of defect characteristics on the I_{DDQ} levels through analytical and experimental results on a simple 3-inverter chain.

*J.A. Segura is at the Balearic Islands University (UIB) and is working on his Ph.D. at the Polytechnical University of Catalonia.

The bridging defects parameter studied was the bridge resistance. In gate oxide shorts two defects parameters were considered: the geometrical location in the defective transistor and the series resistance of the short. Finally, floating transistor gate defects due to *large* opens in the polysilicon conductor controlling the gate of a transistor were studied. The influence of the location of the break as well as the effect of overlapping metal lines was considered.

After modeling the defects at electrical level, the 3-inverter circuit was analyzed using the proposed models for a single defect in the middle inverter. The circuit voltage and current behavior were characterized in a wide range of situations.

Experimental data have been gathered from circuits with defects included by design, showing acceptable agreement with the simulations based on the proposed models.

2. Quiescent Current Estimation and Experimentation

2.1. Bridging Defects

2.1.1. Electrical Level Model. A *bridging defect* appears when two or more lines are, in an undesired way, electrically connected within an integrated circuit. Two types of bridges, vertical and horizontal, can appear depending on the layers connected by the defect [8].

The bridging defect, typically, cannot be modeled by the stuck-at model approach, since, a bridge often does not behave as a permanent stuck node to a logic value [9]–[12]. The model used for bridges is presented in [8] consisting in a resistance connecting the two bridged nodes.

2.1.2. I_{DDQ} Analysis of a Defective Circuit. We evaluated current testing in CMOS circuits in the presence of bridging defects using a simple 3-inverter chain. The bridging defects considered are those affecting the nodes of the middle inverter as shown in Figure 1.

To test the defective circuit with the I_{DDQ} approach, the two bridged nodes must be driven to opposite logic states. Let us analyze a typical bridge (S1) in detail. To detect this defect an input pattern must drive node V_2 to the logic low value, as this node is assumed to be bridged with the power rail (logic high value). Thus, a path from power to ground appears allowing the existence of an abnormal high current (Figure 2). Likewise, for the other defects.

In this section the I_{DDQ} values for the 3-inverter chain are characterized. Several inverters with different relative areas will be considered. Thus, while the size

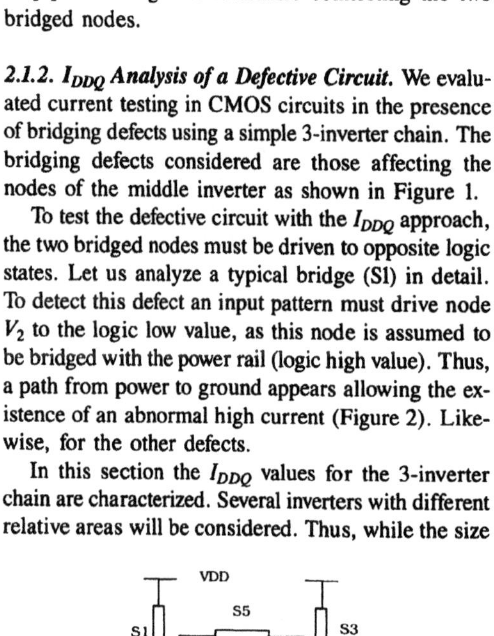

Fig. 1. Bridges (S1-S5) considered in the defective circuit.

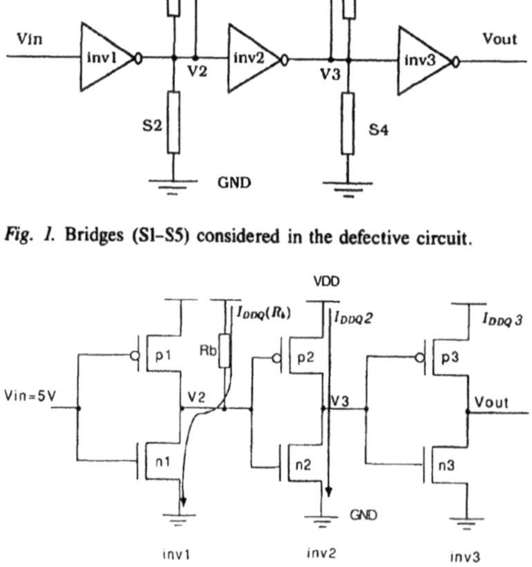

Fig. 2. I_{DDQ} path from power to ground for S1.

of the second and third inverters is constant, the size of the first inverter will be changed. To quantify this ratio, a parameter K is introduced. This parameter K is defined as the ratio of channel width between the first and second (and third) as follows:

$$K = \frac{W_{p2}}{W_{p1}} = \frac{W_{n2}}{W_{n1}}$$

$$W_{n2} = W_{n3} = W_n \quad \text{and} \quad W_{p2} = W_{p3} = W_p$$

The results for the I_{DDQ} dependence on the K parameter and R_b values are presented in Figure 3 where the I_{DDQ} values are classified for (R_b, K) pairs going from (0 kΩ, 1/3) to (100 kΩ, 3) values and, also, the boundary logic testing line is depicted. This line represents the R_b value, for each K parameter value, from where the circuit is logically correct. The results show that the I_{DDQ} dependence on the relative size is high for the small R_b range. This important influence found between I_{DDQ} values and the inverter's relative size is due to the fact that I_{DDQ} values are directly dependent on the resistance offered by the transistor in the conducting path. This resistance is dependent on the W and L values. Otherwise, values of R_b larger than 50 kΩ result in an I_{DDQ} dependence. In this case the I_{DDQ} dependence on the resistance offered by the transistors in the conducting path, still exists, but, the effect of R_b on the total resistance is more important.

For a wide range of R_b, from zero to resistances on the order of a hundred kΩ, the value of I_{DDQ} changes significantly from more than 2 mA to 50 μA. Thus, bridges, one of the most frequent defects in these IC CMOS technologies, seem to exhibit a very good current testability. A critical resistance value is found, below which the circuit produces a logic error and high quiescent current (resistance values around 3.5–4 kΩ). In this domain the circuit is logically as well as current testable. Above the critical resistance the circuit produces logically correct voltages, however, the current consumed remains significantly higher than the current consumed by the defect-free circuit. In this region the circuit is current testable but not logically testable. Delay testing might be also possible in this region. Weakened voltage values, logically correct, might produce detectable delay faults.

2.1.3. Experimental Results. To study the effect of the bridge characteristics (resistance of the bridge), a circuit with a bridge S1 (output bridged to power) was designed corresponding to K = 1 (Figure 4) with the ES2's Dual Layer Metal, 2 μm, n-Well CMOS technology. In this

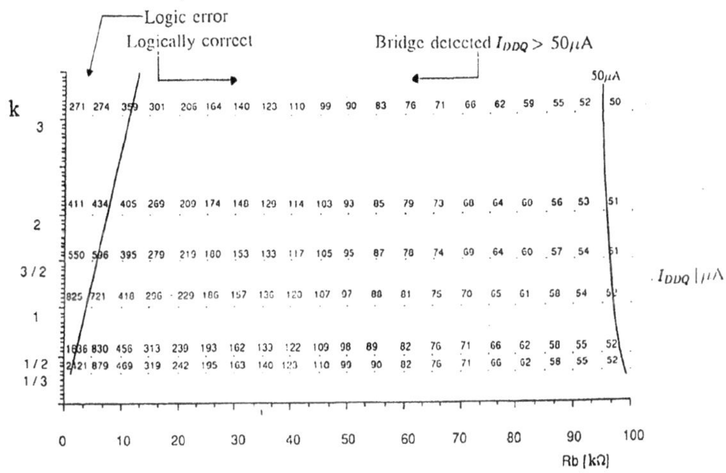

Fig. 3. I_{DDQ} dependence on K, R_b for S1.

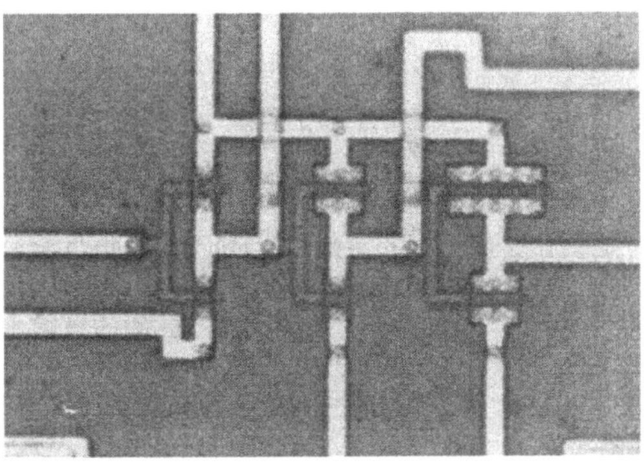

Fig. 4. Photograph of one defective circuit.

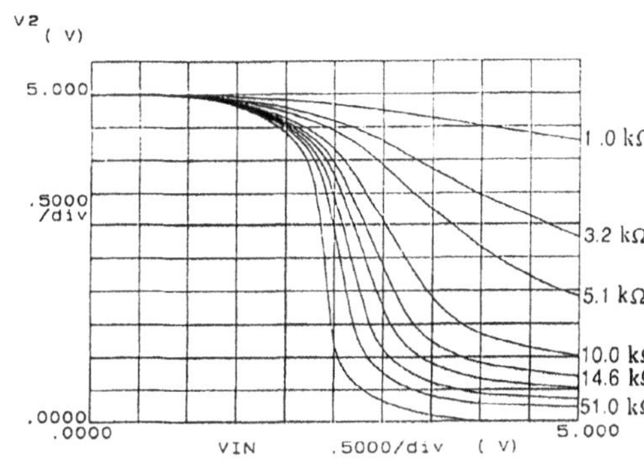

Fig. 5. V_2 vs. V_{in} experimental results (Figure 1).

circuit the defect model was introduced by designing two pads that could externally connect the resistance R_b. In Figure 5 the first inverter outputs for a set of R_b ranging from 1 kΩ to 51 kΩ are presented versus V_{in} voltage. We see, for R_b below 5 kΩ the voltage response of the defective inverter appears logically incorrect, however, for resistances higher than this value the weakened parametric response can be logically recovered along the forward propagation. In Figure 6 the corresponding current consumptions I_{DDQ} (R_b) as V_{in} changes from 0 to 5 V are depicted. It can be noted that the largest deviation is obtained for V_{in} at 5 V when the bridged nodes are driven to opposite states. Figure 7 shows experimental values (Figures 5 and 6) placed SPICE simulated curves (I_{DDQR_b}, I_{DDQ2}, I_{DDQ3} and I_{DDQ}) vs. R_b value. The agreement between simulated and experimental measures of I_{DDQ} is good. It can be noted that the maximum value of I_{DDQ} is obtained at a resistance of the bridge of approximately 3.5 kΩ and not for R_b = 0 Ω. This can be explained by the current contribution of the nondefective stages (I_{DDQ2} and I_{DDQ3}).

The experimental measures of the defective circuits show acceptable agreement with the simulated values. The experimental values obtained present a maximum deviation of 8.5% of the total I_{DDQ} for low resistive shorts. For high resistances of the bridge, agreement is very good with deviations below 0.1%. This behavior has been observed for all the bridging defects of the 3-inverter chain.

2.1.4. Discussion. From the results presented in this subsection, I_{DDQ} current testing appears as an adequate technique for bridging failures. For values of the

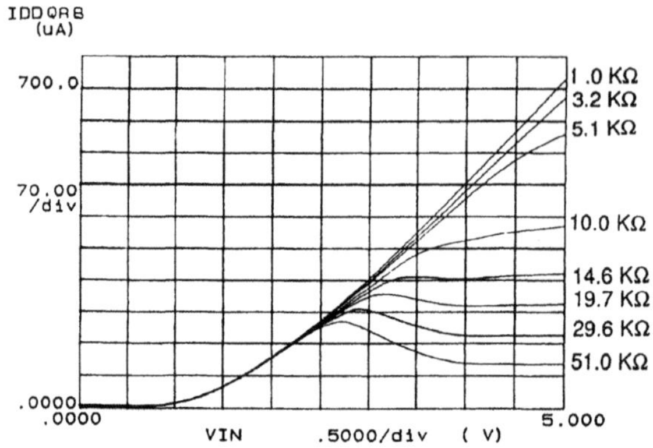

Fig. 6. Current consumption ($I_{DDQ}(R_b)$) vs. V_{in} experimental results for different R_b values, (SI).

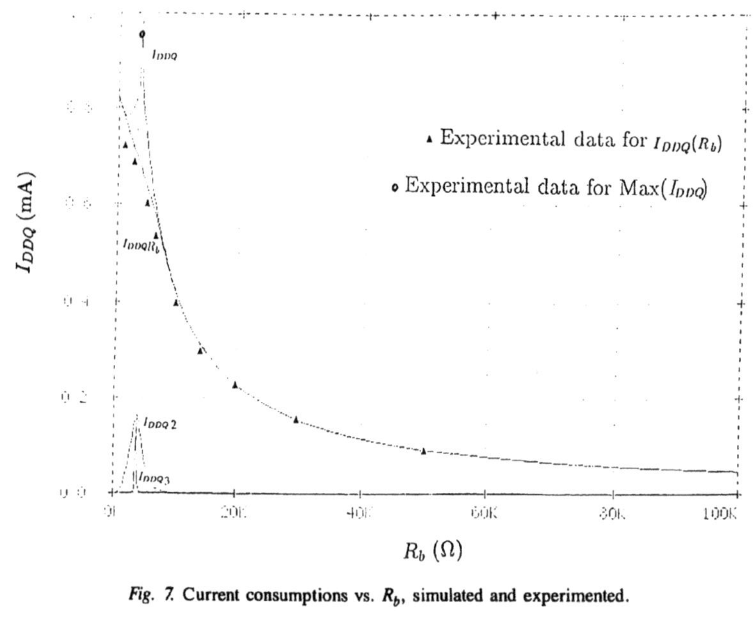

Fig. 7. Current consumptions vs. R_b, simulated and experimented.

bridge resistance lower than a critical value the failure is detectable both by logical and current testing techniques. On the other hand for higher values the circuit gives a logically correct response but the current levels of the circuit still allow potential current testing.

2.2. Gate Oxide Short Defects

2.2.1. The Gate Oxide Short Failure. Gate oxide short (GOS) failures appear frequently in CMOS technology [13]. For these failures an undesired path of current through the oxide of the gate appears in MOS transistors, thus the defective device presents a violation of the gate isolation principle of MOS technology.

GOS defects are short-circuits between the gate electrode and the active zone through the SiO_2 oxide of the device. Figure 8 shows a section of a gate oxide short defective *n*-channel transistor. The short is represented in the figure as an extension of the gate electrode to the transistor channel. In this approach only GOS defects on the channel zone are considered, ignoring shorts between the gate and drain or source diffusion zones.

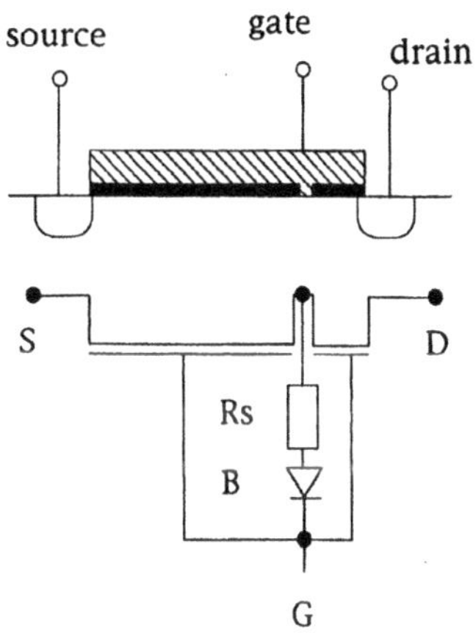

Fig. 8. Circuit level gate oxide short defect model.

An unidimensional circuit level defect model was presented in [14]. A defective MOS transistor is modeled by three components, a rectifying barrier B and two minor transistors in which the GOS defect splits the original transistor.

The barrier B models the rectifying behavior of the new current path introduced by the defect. The relative size of the minor transistors depends on the defect location on the channel.

Hence, the parameters of the model are k and R_s. The first parameter models the place where the defect is located and the second one the effective resistance of the short.

This model satisfies the experimental behaviors of GOS faulty transistors presented in [13] and because of the unidimensionality assumption of the model its tractability at electrical level is enhanced. In [15] a bidimensional model of the defect can be found.

In the next subsection the model is applied to a 3-inverter chain circuit whose middle inverter n-MOS transistor is assumed defective.

2.2.2. *Modeling of the I_{DDQ} of a Defective Circuit.*

Figure 9 shows the defective circuit. The behavior of this circuit was investigated by SPICE simulation using the circuit level defect model presented in Section 2.2.1.

The I_{DDQ} values of the chain circuit are shown in the curves of Figure 10 considering the parameters of the model R_s and k. The value of the current is expressed in μA (in a defect-free circuit this quiescent current is in the order of tens of nA) showing that the defect effect on I_{DDQ} is stronger when the defect is located near the source.

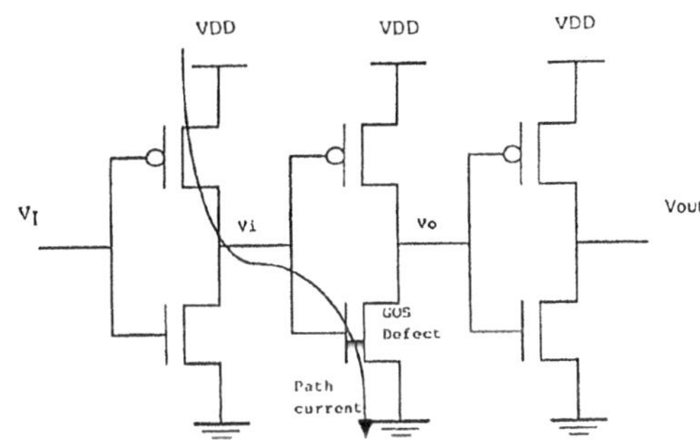

Fig. 9. Inverters chain with a GOS.

For all the defective points on the plane of Figure 10, I_{DDQ} presents a value 3 or 4 orders of magnitude higher than defect-free circuits, so the defects are current testable. When R_s increases, the value of I_{DDQ} decreases, implying that only for very high values of R_s (in the range of MΩ) the defect cannot be detected by I_{DDQ} testing. However, the area of defects detected by logical testing is very reduced as is indicated in the dotted area domain. This means that GOS defective circuits may exhibit an error free logical behavior even for defects producing a significant level of I_{DDQ}. Figure 11 shows the voltage response of the 3-inverter chain for GOS resistances of 50 kΩ and 50 Ω. For R_s = 50 kΩ the logical behavior is correct. For R_s = 50 Ω, the analog deviation is high enough to produce logic error (Figure 10).

So we conclude that logical testing has a limited ability to detect GOS defects even when the defect produces serious I_{DDQ} current deviations. The I_{DDQ} defect detection domain covers the whole area of logic testing.

2.2.3. *Experimental Results.*

Experimental results were obtained from designed GOS defective MOS circuits. By using VLSI design facilities, defective MOS transistors were obtained by shorting the gate electrode to the channel of the device. The methodology was described in [14]. Figure 12 shows a photograph of a defective designed device.

Following this methodology a chain with 3-inverters was designed and fabricated using a standard CMOS process (ES2's Dual Layer Metal, 2 μm, n-Well CMOS technology).

Figure 13 shows the I_D (global I_{DDQ} of the chain) versus V_I (input voltage at first stage) relationship for three different values of the k parameter. The fault behavior appears when V_I is low in all the cases

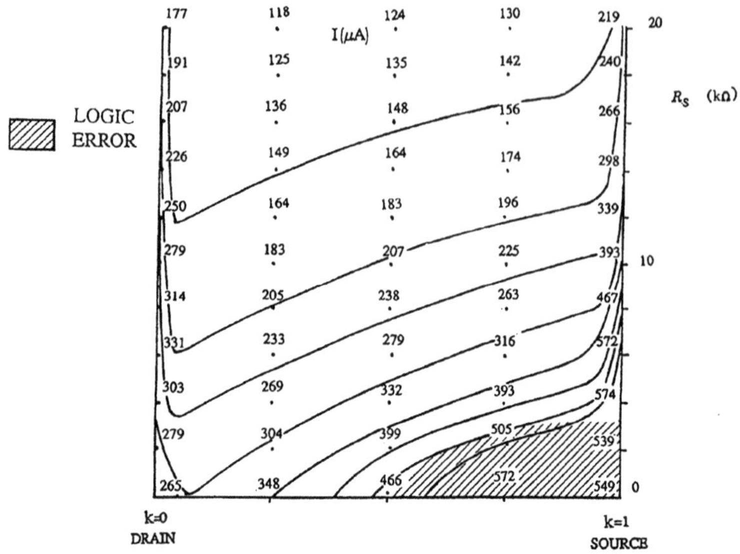

Fig. 10. Logic and I_{DDQ} testing zones in function of the parameters of the model k and R_s.

because the defective gate current increases. The curve which presents higher current values corresponds to the case in which the defect is closer to the transistor's source ($k = 0.9$). When the defect is closer to the drain the leakage current is lower ($k = 0.5$ and $k = 0.1$).

Figure 14 indicates the behavior of the input and output voltage at the defective stage for the same V_I range. This behavior and the levels of I_{DDQ} measured by experimentation agree adequately with that predicted by the model.

2.2.4. Discussion. I_{DDQ} levels of CMOS circuits are sensitive to the presence of GOS defective devices. A simple electrical model can be used to evaluate abnormal current, voltage behavior and testing strategies. Results from the modeling agree with results obtained from experimentation using design-based defective devices.

It also has been observed that higher current values are obtained when the defect is closer to the transistor's source.

2.3. Floating Gate Defect

2.3.1. Electrical Level Model. The class of opens considered in this work is restricted to those causing a transistor gate to float [4], [7], [16]. In other words, the poly path is broken and the electrical connectivity to the gate is lost. We consider large breaks so that tunneling effects (studied by Henderson et al. [6]) in the break can be assumed negligible for this class of opens. In our studies the floating gate gets a voltage determined by the capacitive coupling of the broken poly path to the metal lines crossing it. The treatment of this defect as a *stuck-open transistor* [17] is shown to be inadequate in some defective circuits defects as reported previously [4], [7], [16]. The work presented here shows that for certain *floating gate defects* the transistor may conduct and for others it may remain stuck-open. A modeling taking into account the poly-bulk and metal-poly capacitances allows a characterization at electrical level of the defect and its influence on the quiescent current.

We have modeled the break defect by two capacitances: the capacitance poly-bulk capacitance (C_{pb}), that depends on the poly track from the gate to the break, and the capacitance metal-poly capacitance (C_{mp}) that depends on the overlapped metal-poly (floating) tracks (Figure 15).

The gate voltage on the floating gate transistor depends on the coupling capacitances to the gate and of the voltages at each adjacent coupled node [4], [7],

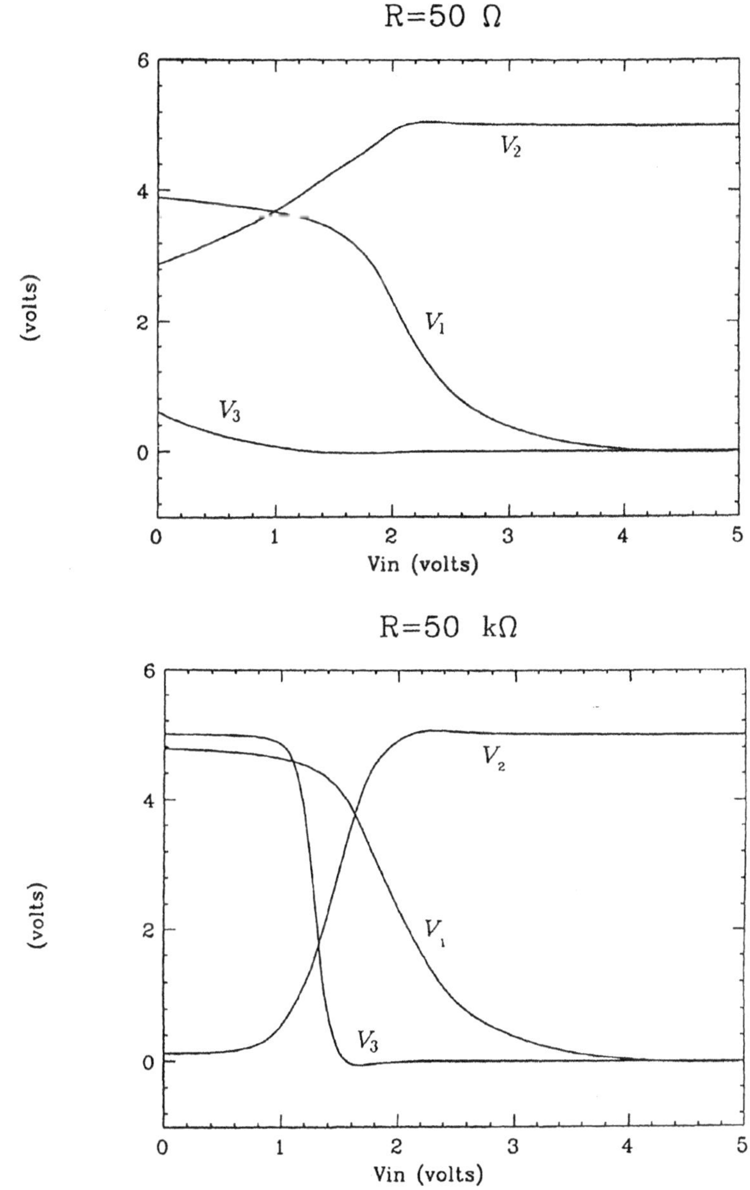

Fig. 11. Logic behavior of the chain for defect contact resistance $R_s = 50$ kΩ and $R_s = 50$ Ω.

[16]. We used the charge-oriented model of Ward and Dutton [18] implemented in the SPICE electrical simulator. The capacitances values were calculated from the circuit layout measurements taking into account the technology parameters.

2.3.2. I_{DDQ} Analysis of a Defective Circuit. Consider a CMOS static 3-inverter chain with a large break in the poly path controlling a transistor gate. We assume a break defect at the gate of the *n*-channel transistor at the middle inverter and the floating gate is influenced

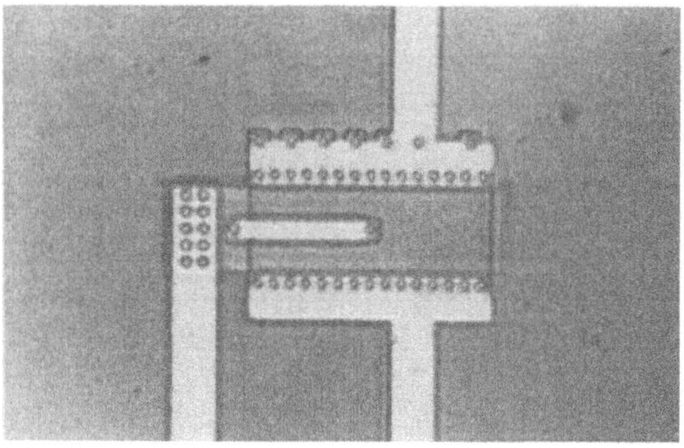

Fig. 12. Photograph of a GOS defective designed MOS transistor.

by a controllable signal from a level of metal track (VMET). The electrical model of the defective circuit is indicated in Figure 16. It is assumed that the behavior of the defective transistor depends on the coupling capacitances previously mentioned.

In order to get enough current through the defective transistor, the PMOS transistor of the middle inverter must be *on*. A path from the power supply to *GND exists if the n*-channel transistor gate gets enough voltage to turn it on so that an abnormal current increment of the I_{DDQ} can be sensed. The simulation results show the voltage induced at the floating gate is enough to turn on the faulty transistor in a wide range of realistic

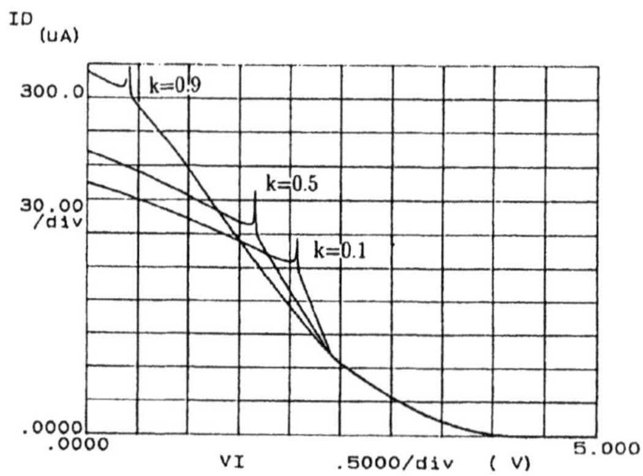

Fig. 13. Global current versus input voltage for a 3-inverter chain with a GOS defect at the middle inverter (n-MOS transistor). The three curves correspond to three different locations of the defect (k = 0.9, 0.5 and 0.1).

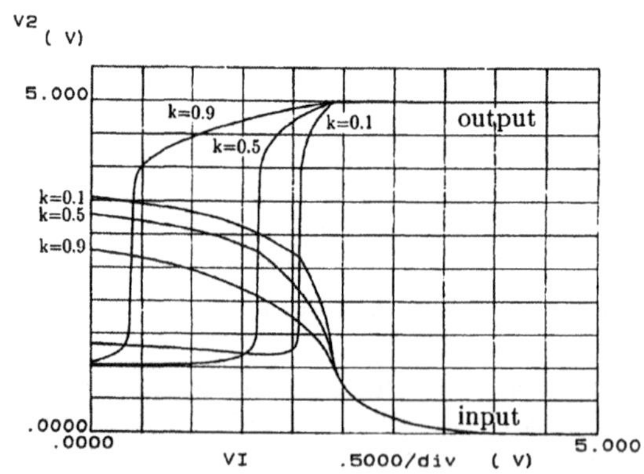

Fig. 14. Input and output voltages of the defective inverter versus input voltage of the chain. The different curves correspond to three different values of the k parameter.

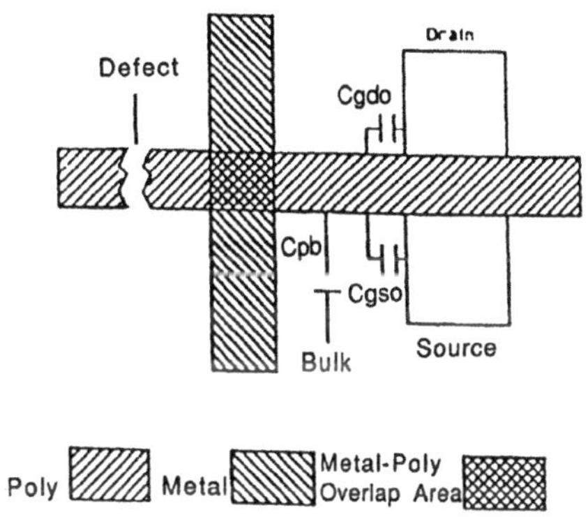

Fig. 15. Circuit layout of the floating gate transistor.

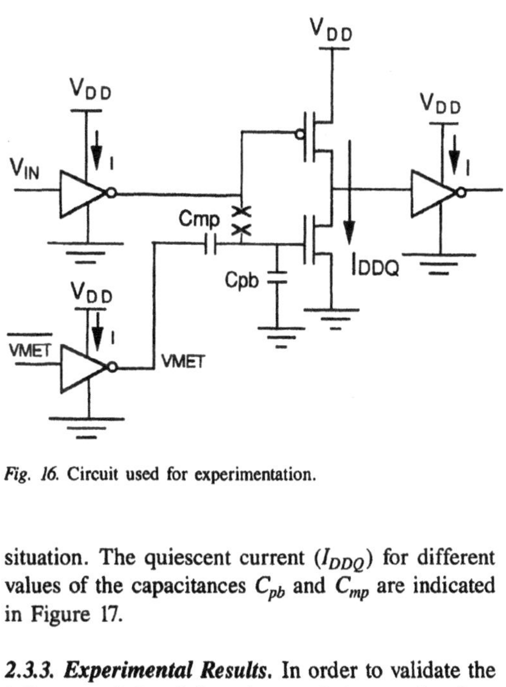

Fig. 16. Circuit used for experimentation.

situation. The quiescent current (I_{DDQ}) for different values of the capacitances C_{pb} and C_{mp} are indicated in Figure 17.

2.3.3. Experimental Results. In order to validate the influence of the defect characteristics on the I_{DDQ} levels, five defective circuits (inverter chains) with a floating gate defect were designed and fabricated using a standard CMOS process (ES2's Dual Layer Metal, 2 μm, n-Well CMOS Technology). In each circuit different defects affecting the *n*-channel transistor of the middle inverter were intentionally constructed with different lengths (from 3 to 15 μm) of the unconnected poly. The values of C_{pb} change from circuit to circuit ranging from 1.95 fF to 5.3 fF. To analyze the influence of the metal by parasitic metal-poly capacitive coupling, different areas of metal-poly overlap were designed. The values of C_{mp} ranges from 1.1 fF to 5.4 fF.

A photograph taken with an electronic microscope of a designed floating gate *n*-channel transistor is shown in Figure 18. The gate of the transistor is not connected to any control signal and the floating gate is influenced through a controlled signal in the metal track. The signal of the metal track is driven by an inverter with a low level voltage at its input.

2.3.4. Analysis of Results. The experimental I_{DDQ} data for the inverters chain were plotted in the C_{pb}-C_{mp} plane (Figure 19). The results show a good qualitative agreement with the simulation. The experimental values present good agreement with the simulated I_{DDQ} for low I_{DDQ} values. However, the predictions of the SPICE simulations are higher than the experimental values at high I_{DDQ} (see Figures 17–19). This discrepancy may be explained taking into account the imprecision in estimating the C_{pb} and C_{mp} values of the experimental circuits. Nevertheless, the fact that the levels of I_{DDQ} increase for higher influence (more overlap metal-poly) are clearly consistent experimentally and by simulated analysis. Similarly, the higher C_{pb} (more area of poly floating) decreases the I_{DDQ} levels.

In order the investigate the influence of the analog input voltages on the quiescent current, the static transfer characteristics for the five defective circuits were obtained (Figure 20). The current *IGND* is the total current returning to ground.

The experimental results show that the values of I_{DDQ} present a maximum for intermediate voltages due to the contribution of the nondefective inverters and a remnant I_{DDQ} for logical input values. In this situation, in which the PMOS transistor of the defective inverter has low voltage at its gate (very close to 0 V) and the NMOS transistor is influenced by an overlapping metal track at 5 V, sufficient I_{DDQ} levels for current test detection are produced. If the metal is not influencing (VMET = 0 V) the floating gate transistor may operate at subthreshold and the defect would not be current testable.

3. Conclusion

The quiescent current consumed by a circuit was found to be sensitive to the presence of the defects considered in the work: bridges, GOS and floating gates.

The levels of I_{DDQ} caused by these defects are dependent on the defect characteristics. Consequently the

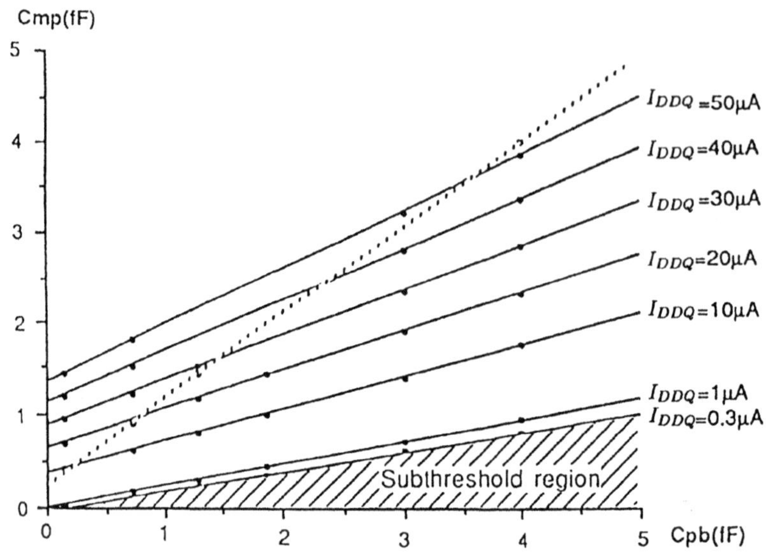

Fig. 17. I_{DDQ} curves for the defective inverters in the C_{pb}-C_{mp} plane V_{IN} = 5 V and VMET = 5 V.

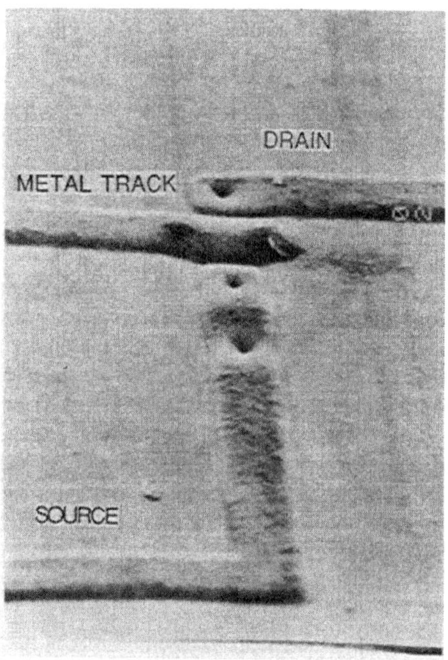

Fig. 18. Photograph of a designed defective floating gate transistor. The metal track overlaps the poly track.

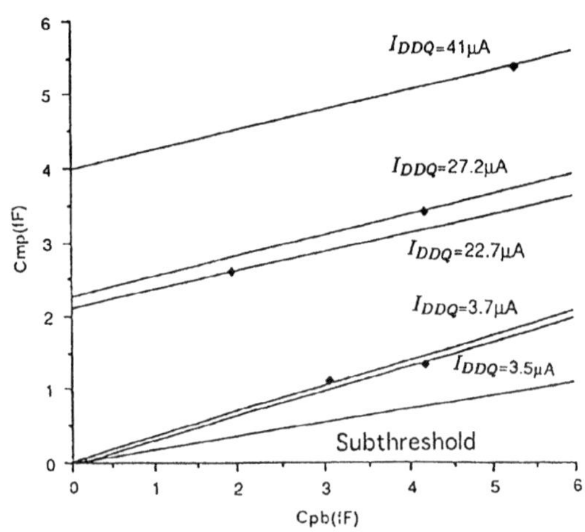

Fig. 19. I_{DDQ} experimental measures for the defective inverters chain in the C_{pb}-C_{mp} plane V_{IN} = 5 V and VMET = 5 V.

main parameters influencing I_{DDQ} have been investigated, through analytical simulations and experimental measures of circuits designed with different defect parameters. Analytical models at electrical circuit level have been proposed and used to estimate the influence of each defect parameter. Circuits designed with intentional defects have been used to experimentally characterize the I_{DDQ} sensitivity to each defect parameter. Experimental and simulated data show acceptable agreement and permit to investigate the testability by current or voltage of the studied defects.

Current testing appears as an adequate technique for bridge failures. The resistance of the bridge has been taken into account in the modeling and analysis. From

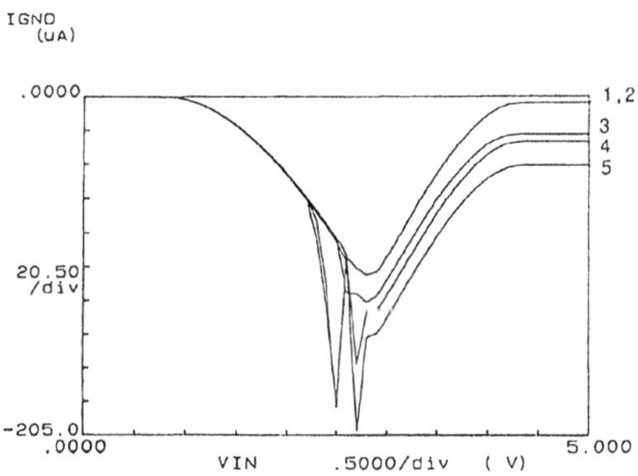

Fig. 20. Transfer characteristics of the I_{DDQ} for five different defective circuits

1. $C_{pb} = 3\,fF$ $C_{mp} = 1.1\,fF$
2. $C_{pb} = 4.17\,fF$ $C_{mp} = 1.34\,fF$
3. $C_{pb} = 1.95\,fF$ $C_{mp} = 2.6\,fF$
4. $C_{pb} = 4.17\,fF$ $C_{mp} = 3.42\,fF$
5. $C_{pb} = 5.28\,fF$ $C_{mp} = 5.4\,fF$

that, a critical resistance value was found for the circuit behavior evaluation. For values of the bridge resistance lower than this critical resistance, the failure is detectable both by logical and current testing techniques. On the other hand, for higher values of resistance the circuit gives a logically correct response but the current levels of the circuit still allow potential current testing. Evidence is shown on the limited capability of the logic classical testing for bridge defects and the interest of quiescent current testing.

Experimental data show good agreement with simulations and corroborate the dependence of I_{DDQ} on the bridge resistance. For some bridges it has been shown analytically and experimentally that the maximum I_{DDQ} is not obtained at zero resistance.

For GOS defects, a characterization of the possible parameters of the short in the plane (k, R_s) is given. The parameter k is the localization of the short in the channel and R_s is the effective short resistance. It is shown that the domain of current testability is larger than the domain of logic testability. In most cases, the defect produces intermediate voltage levels which may be interpreted as logically correct, however the quiescent current is orders of magnitude higher than the current consumed by the defect-free circuit.

Experimental data obtained from intentionally designed GOS defective circuits agree quantitatively in an acceptable way with the results predicted by the modeling and qualitatively with the experimental data obtained by other researchers. Thus the approach used for designed GOS defects seems to be adequate with interesting capabilities to control the parameters of the experimentation.

The results for floating gate large opens show that although the widely used stuck-open transistor fault model can be acceptable for failures that bring the faulty transistor to operate in the subthreshold region, it is inadequate for a wide range of realistic situations. The capacitive coupling may induce enough voltage at the floating gate to force the operation of the defective transistor above threshold creating a channel and consequently a current path for the quiescent current. The defect is characterized in the plane (capacitance metal-poly, capacitance poly-bulk) for a circuit layout in a given technology.

Experimental data on defective circuits with different locations of the break have been obtained in order to investigate the influence of the length of the disconnected poly (C_{pb}) and the capacitive coupling due to metal tracks over the poly (C_{mp}). The measures have shown acceptable agreement for low I_{DDQ} values and consistent qualitative behavior for high I_{DDQ}.

For this defect it can be concluded that if the disconnected poly path is larger (larger C_{pb}) the I_{DDQ} becomes smaller. On the other hand a capacitive influence of a metal track increases I_{DDQ} levels of the defective circuit.

Acknowledgment

This work has been partially supported by the Project EVEREST (ESPRIT Program Project No. 2318) and by the Comisión Interministerial para la Ciencia y Tecnología (C.I.C.Y.T.) project No. TIC 1024/90.

References

1. J.P. Shen, W. Maly, and F.J. Ferguson, "Inductive fault analysis of MOS integrated circuits," *Design and Test of Computer Manufacturing Testing*, pp. 13–26, December 1985.
2. J.P. Teixeira, F.M. Gonçalves, J.J.T. Sousa, "Realistic fault list generation for physical testability assessment," *Workshop on Defect and Fault Tolerance in VLSI Systems*, Grenoble, November, 1990.
3. T. Storey, W. Maly, J. Andrews, and M. Miske, "Stuck fault and current testing comparison using CMOS chip test," *1991 Int. Test Conf.*, pp. 311–318, October 1991.
4. W. Maly, P.K. Nag, and P. Nigh, "Testing oriented analysis of CMOS ICs with opens," *Proc. of ICCAD*, pp. 344–347, 1988.
5. J.M. Soden, R.R. Fritzemeier, and C.F. Hawkins, "Zero defects or zero stuck-at faults—CMOS IC process improvement with I_{DDQ}," *Int. Test Conf.*, pp. 255–256, September 1990.
6. C.L. Henderson, J.M. Soden, and C.F. Hawkins, "The behavior and testing implications of CMOS IC logic gate open circuits," *Int. Test Conf.*, pp. 302–310, October 1991.

7. M. Renovell and G. Cambon, "Topology dependence of floating gate faults in MOS circuits," *Electronics Letters*, vol. 22, pp. 152–153, January 1986.
8. R. Rodríguez-Montañés, J.A. Segura, V.H. Champac, J. Figuera, and J.A. Rubio, "Current vs. logic testing of gate oxide short, floating gate and bridging failures in CMOS," *Int. Test Conf.*, pp. 510–519, Nashville, TN, 1991.
9. J.M. Acken, "Testing for bridging faults (shorts) in CMOS circuits," *20th Design Automation Conf.*, pp. 717–718, 1983.
10. Y.K. Malaiya, A.P. Jayasumana, and R. Rajsuman, "A detailed examination of bridging faults," *Int. Conf. on Comp. Design*, pp. 78–81, 1986.
11. J.M. Soden and Ch.F. Hawkins, "Electrical properties and detection methods for CMOS defects," *Proceedings of the 1st European Test Conf.*, pp. 159–167, April 1989.
12. H. Hao, and E.J. McCluskey, "Resistive shorts within CMOS gates," *Int. Test Conf.*, pp. 292–301, Nashville, TN, 1991.
13. C.F. Hawkins, and J.M. Soden, "Electrical characteristics and testing considerations for gate oxide shorts in CMOS ICs," *Int. Test Conf.*, pp. 544–555, November 1985.
14. J.A. Segura, A. Rubio, and J. Figueras, "Analysis and modeling of MOS devices with gate oxide shorts failures," *Int. Symp. on Circuits and Systems*, pp. 2164–2167, June 1991.
15. M. Syrzycki, "Modeling gate oxide shorts in MOS transistors," *IEEE Transactions on Comp. Aided Design of IC and Systems*, vol. 8, pp. 193–202, March 1989.
16. V.H. Champac, R. Rodrígues-Montañés, J.A. Segura, J. Figueras, and J.A. Rubio, "Fault modeling of gate oxide short, floating gate and bridging failures in CMOS circuits," *European Test Conf.*, pp. 143–148, 1991.
17. R. Chandramouli, "On testing stuck-open faults," *13th Fault Tolerant Computing Systems*, 1983.
18. D.E. Ward and R.W. Dutton, "A charge-oriented model for MOS transient capacitances," *J. Solid-State Circuits*, pp. 703–707, October 1978.

Jaume A. Segura obtained his M.S. degree from the Physics Department of Balearic Islands University.

He has been an associate professor of Electronics of the Elecronic Engineering Department at the Industrial Engineering Faculty (UPC). He is currently doing his Ph.D. in the area of electronic technology with a grant from the Spanish Government. His research interests include device and circuit modeling, VLSI design and test and high speed circuit design.

Víctor Champac graduated in electronic engineering at the Universidad Autónoma de Nuevo León, México.

He is working toward the Ph.D. degree in electronic technology at the Politechnical University of Catalonia, Spain. His research interests include modeling of defects and testing of CMOS VLSI circuits. In particular he is actually working on testing of floating gate defects.

Rosa Rodrígez obtained her M.S. degree from the Physics Department of the Universidad de Barcelona.

She is currently an associate professor of the Electronic Engineering Department at Industrial Engineering Faculty (UPC) and is working for her Ph.D. in the area of electronic technology. Her research interests include VLSI design, test and circuit modeling.

Joan Figueras received his Industrial Engineering (Electrical) degree from the Universitat Politècnica de Catalunya (U.P.C.) and the Ms.C. (Electrical) and Ph.D. (Computer Information and Control) degrees from the University of Michigan were he was a Fulbright Scholar and worked in the Systems Engineering Laboratory (1968–71). He has been visiting professor in UCLA (1979) and UCI (1992). Since 1983 is Professor at the Electronics Dept. of th U.P.C. with teaching and research responsibilities in the area of Digital Systems and VLSI design. His current research interests include Test and Design for Testability, Defect and Fault Models and new Testing and Fault Tolerance methodologies.

Antonio Rubio obtained his M.S. and Ph.D. degrees from the Industrial Engineering Faculty of Barcelona. He has been associate professor of the Electronic Engineering Department at the Industrial Engineering Faculty (UPC) and professor of Electronics at the Department of Physics, University of Balearic Islands, Spain. He is currently full professor of Electronic Technology at the Telecommunication Engineering Faculty, Electronic Engineering Department, Polytechnical University of Catalunya in Barcelona. His research interests include VLSI design and test, computer-aided design, automatic test pattern generation, device and circuit modeling and high speed circuit design and analysis.

JOURNAL OF ELECTRONIC TESTING: Theory and Applications, 3, 349-357 (1992)

QUIETEST: A Methodology for Selecting I_{DDQ} Test Vectors

WEIWEI MAO AND RAVI K. GULATI
Ford Microelectronics, Inc. 9965 Federal Drive, Colorado Springs, CO 80921-3698

Received August 14, 1992. Revised August 24, 1992.
Editor: V.D. Agrawal

Abstract. Even high stuck-at fault coverage manufacturing test programs cannot assure high quality for CMOS VLSI circuits. Measurement of quiescent power supply current (I_{DDQ}) is a means of improving quality and reliability by detecting many defects that do not have appropriate representation in the stuck-at fault model. Since each I_{DDQ} measurement takes significant time, a hierarchical fault analysis methodology is proposed for selecting a small subset of production test vectors for I_{DDQ} measurements. A software system QUIETEST has been developed on the basis of this methodology. For two VLSI circuits QUIETEST selected less than 1% of production test vectors for covering all modeled faults that would have been covered by I_{DDQ} measurement for all of the vectors. The fault models include leakage faults and weak faults for representing defects such as gate oxide shorts and certain opens.

Keywords: Leakage faults, I_{DDQ} testing, weak faults.

1. Introduction

Manufacturing test programs with even 100% stuck-at fault coverage cannot assure high levels of quality in CMOS VLSI circuits. This is because many common CMOS physical defects such as gate oxide shorts [1]-[5] and opens [3], [6], [7] do not have appropriate representation in the stuck-at fault model. Measurement of quiescent power supply current (I_{DDQ}) [2], [8], [9], [10] has been established as an effective means of improving the quality and reliability of CMOS circuits beyond what can be assured by stuck-at fault model based logic testing.

A direct approach for I_{DDQ} testing requires I_{DDQ} measurement upon the application of every functional/structural logic test vector [10]. However, this is not feasible in large volume production testing because I_{DDQ} measurements at the required levels of sensitivity are time consuming [11], [12].

In this article, we present a methodology for selecting a small number of vectors from the production test set for I_{DDQ} measurements to detect leakage faults [13] and weak faults [14] (Section 2). Production testing could still be done at normal speed except for slowing down as and when a selected I_{DDQ} vector was applied. Our approach conserves test time by combining logic and I_{DDQ} testing without requiring I_{DDQ} measurement for every vector. The methodology is essentially a hierarchical simulation technique and allows switch level logic state information to be inferred from gate level logic simulations. The software implementation QUIETEST has been used to select I_{DDQ} test vectors for CMOS VLSI circuits with very encouraging results.

The article is organized as follows. The target fault models for I_{DDQ} testing are presented in Section 2 while the fault detection conditions are given in Section 3. The hierarchical fault analysis methodology is presented in Section 4. Discussions of software implementation and the results are given in Section 5.

2. Fault Models

The defects targeted for detection by I_{DDQ} measurements are modeled as leakage faults and weak faults.

2.1. Leakage Faults

A defect-free MOS transistor has no current between gate and source, gate and drain, or gate and substrate (bulk). The current between source and bulk or drain and bulk should be less than a specified value if bulk is connected to $V_{SS}(V_{DD})$ for NMOS (PMOS) transistors. The current between source and drain should be less than a specified normal value when the transistor is off.

Some defects, such as gate oxide shorts, will cause leakage current between certain terminals of a MOS transistor. For example, there may be leakage between gate and source or gate and drain. We consider six leakage faults per MOS transistor in our leakage fault model:

f_{GS}—the leakage fault between gate and source
f_{GD}—the leakage fault between gate and drain
f_{SD}—the leakage fault between source and drain
f_{BS}—the leakage fault between bulk and source
f_{BD}—the leakage fault between bulk and drain
f_{BG}—the leakage fault between bulk and gate

We assume that transistor leakage faults do not affect the logic values in the circuit. This is typically true when gate-oxide shorts are in nascent stage at the time of production testing. However, if not screened, CMOS circuits with such defects can gradually develop hard faults causing field failures. Our I_{DDQ} test vector selection methodology targets single leakage faults. However, if multiple leakage faults are present, they will also be detected if the selected test vectors cover at least one of these faults.

2.2. Weak Faults

It is well known that while NMOS transistors can pass logic 0 without degradation, they degrade the voltage level for logic 1 [15]. If the threshold voltage of the device is, say, 0.7V, a 5V level gets degraded to (5 − 0.7) = 4.3V. Actually, the voltage degradation is even more due to body effect [15]. The reverse is true for PMOS devices, i.e., they can pass logic 1 without degradation, but degrade (increase) the level of logic 0. For the circuit in Figure 1, circuit simulation (with I2 at logic 1) confirms the above statements (Table 1). While the voltage levels at node A are significantly degraded, the logic values are still interpreted correctly by the following stage. This is the reason such defects cannot be detected by logic testing and are often classified as *undetectable* and excluded from fault list and analysis during fault simulation. However, in the actual device such defects (and resulting degraded voltage levels) can increase propagation delays and reduce noise immunity, and therefore it is important to detect them.

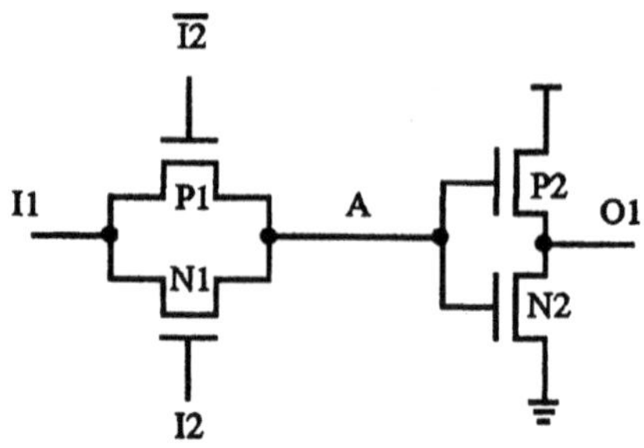

Fig. 1. An example circuit.

Table 1. Circuit simulation results for the example circuit (Figure 1), $V_{I2} = 5V$.

Defect	V_{I1}	V_A	V_{O1}
P1 Open	5.0V	3.6V	0.1V
N1 Open	0.0V	1.5V	4.8V

If a node has a normal logic 0(1) voltage in a defect-free circuit and has a weak (degraded) logic 0(1) voltage in the defective circuit, we model it as a *weak-0(1) fault* at the node. Note that we exclude the case where a node may have a degraded voltage level if it is not driven (driving stage is tristated).

The identification of weak faults is more difficult than leakage faults. We begin with definitions that establish a framework for creating the list of all weak faults in the circuit.

A path between nodes n_i and n_j is called a *transmission path* if there is a chain of channel connected transistors between n_i and n_j. If all transistors in a transmission path are turned on, the path is called a *conducting path*. A transmission path consisting of only PMOS (NMOS) transistors is called a *normal-1(0) transmission path*. A transmission path with at least one PMOS (NMOS) transistor is called a *weak-0(1) transmission path*. A conducting path consisting of only PMOS (NMOS) transistors is called a *normal-1(0) conducting path*. A conducting path with at least one PMOS (NMOS) transistor is called a *weak-0(1) conducting path*.

For example, in Figure 2, transistors N4 and N5 form a normal-0 transmission path between V_{SS} and node A. Transistors P2 and P3 form a normal-1 transmission path between V_{DD} and O1. If I1 = 1 and I2 = 0, transistors P2 and P3 also form a normal-1 conducting path between V_{DD} and O1. For the same inputs, transistors P2 and N3 form a weak-1 conducting path between V_{DD} and O1.

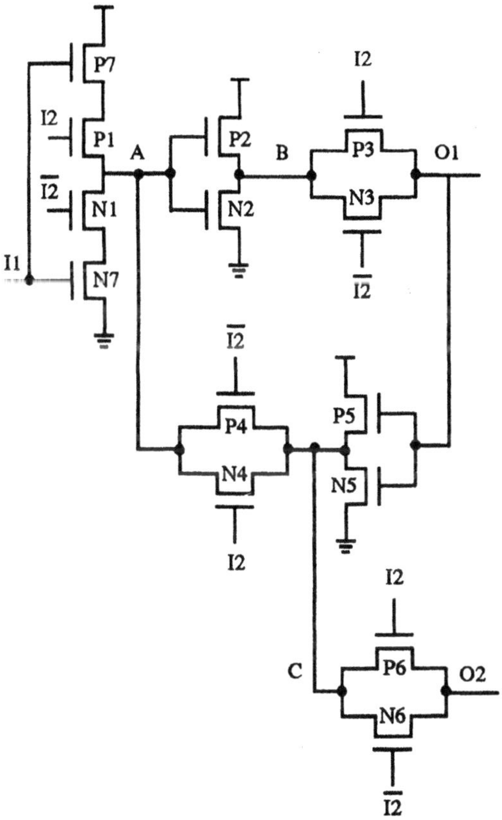

Fig. 2. A circuit with transmission, normal and weak conducting paths.

The *necessary* and *sufficient* conditions for a node n_i to have a weak-0(1) fault are:

(1) there should be at least one normal-0(1) conducting path between n_i and $V_{SS}(V_{DD})$ in the defect-free circuit, and all normal-0(1) conducting paths should be blocked in the defective circuit;

(2) there should be at least one weak-0(1) conducting path between node n_i and $V_{SS}(V_{DD})$ in the defective circuit [14].

We must point out that not all nodes can have weak-0 and weak-1 faults. On the other hand, it is possible for some nodes to have both weak-0 and weak-1 faults. Other nodes may have only weak-0 or only weak-1 faults [14].

It is also important to note that a node may have more than one weak-0(1) faults. For instance, in the circuit in Figure 3, if I1 = 1, I2 = 1, I3 = 0 and I4 = 1, a stuck-at 1 at line I3 will turn P2 off and cause a weak-1 fault at node A. If I1 = 1, I2 = 1, I3 = 1 and I4 = 0, a stuck-at 1 at line I4 will turn P3 off and cause a weak-1 fault at node A. The weak-1 faults at node A caused by different defects are considered as two different weak-1 faults because the detection of one of them cannot guarantee the detection of the other.

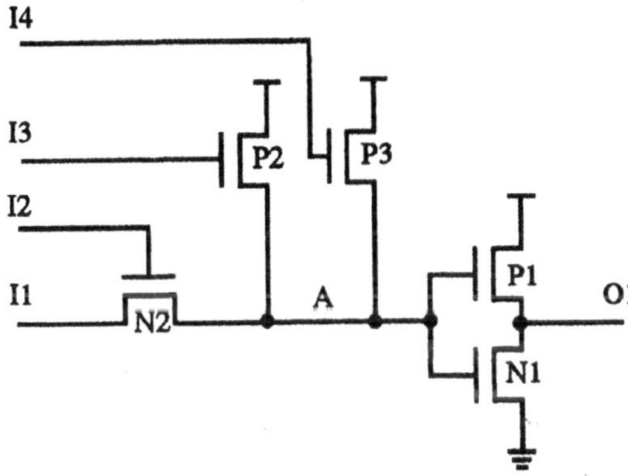

Fig. 3. A circuit with multiple weak faults at node A.

Similar to the single fault assumption widely adopted in stuck-at fault testing we assume that a weak-0(1) fault at a node is caused by blocking of all normal-0(1) conducting paths due to a single defective transistor. For example, in Figure 3, we do not allow the case where node A may have a weak fault when I1 = 1, I2 = 1, I3 = 0 and I4 = 0 and both P2 and P3 are stuck-open.

3. Fault Detection

This section discusses the conditions for detecting leakage faults and weak faults.

3.1. Leakage Faults

The detection of leakage faults depends upon the logic state of the circuit. For example, in Figure 4, assume that transistor N2 has a gate to source leakage fault f_{GS}. This fault will be detected only if node A is at logic 1 (input IN is at logic 0). This creates a leakage current path from V_{DD} to V_{SS} through the fault as shown.

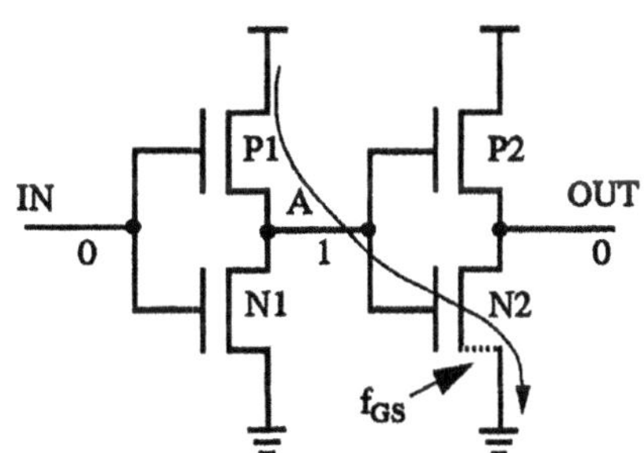

Fig. 4. Detection condition for leakage fault f_{GS} in transistor N2.

It can be inferred that the requirement for detection of leakage faults is that the two transistor terminals with leakage must be at opposite logic levels, and must have driving strengths. Non driving or high impedance logic states are not sufficient because current path from V_{DD} to V_{SS} cannot be created through them. Note that both fault sensitization and fault effect propagation are implicit in this detection requirement.

3.2. Weak Faults

A circuit in which a defect maps into our weak fault model was discussed in [6]. It was reported that if one of the two transistors in a CMOS transmission gate is stuck-open, and if it drives a load inverter, then the fault can be detected with certainty by measuring I_{DDQ} for test vectors that apply logic 0 and logic 1 to the input of the enabled transmission gate. For example, in the circuit of Figure 1, if a logic 1(0) is applied when transistor P1 (N1) is open, it causes degraded logic voltage at the input of the load inverter. That in turn causes both the transistors in the inverter to remain at least partially turned on leading to elevated I_{DDQ}. For this circuit, I_{DDQ} in the presence of the defect increases to at least 56 μA (Table 2) from essentially 0 μA. However, it can be easily seen that if the transmission gate feeds one of the two inputs of a NAND gate (Figure 5) instead of an inverter, similar faults in the transmission gate

Table 2. Circuit simulation results for the example circuit (Figure 1), V_{I2} = 5V.

Defect	V_{I1}	V_A	I_{DDQ}
P1 Open	5.0V	3.6V	56 μA
N1 Open	0.0V	1.5V	94 μA

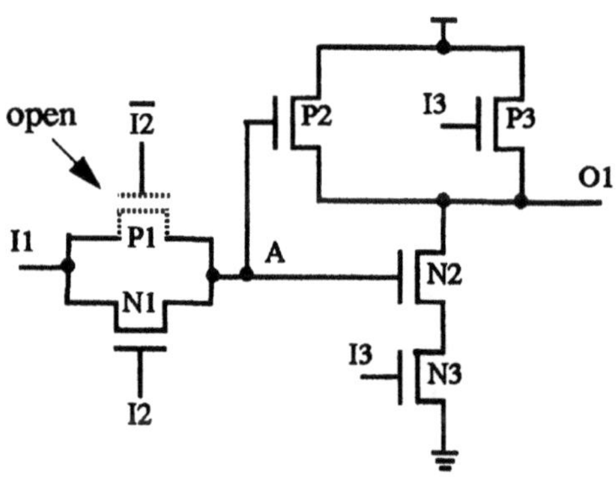

Fig. 5. The sensitization and propagation of a weak-1 fault at node A.

will not be detected by unconditionally applying both logic 0 and 1 at the transmission gate input. Although transistors P2 and N2 in the NAND gate will both be at least partially turned on by the degraded logic voltage at node A, an I_{DDQ} path from V_{DD} to V_{SS} cannot be created unless the other input of the NAND gate (node I3) is also driven to a logic 1. Thus, in general, the circuit state for I_{DDQ} measurement needs further consideration and unlike leakage faults, weak fault sensitization and its effect propagation must be separately addressed.

3.2.1. Sensitization of Weak-0(1) Faults. A weak-0(1) fault at a node is sensitized by a test vector if due to a defective transistor all normal-0(1) conducting paths to the node are disconnected but at least one weak-0(1) conducting path is maintained. For example, in Figure 5, if transistor P1 cannot conduct due to a defect, the weak-1 fault at node A will be sensitized if I1 = 1 and I2 = 1. This is because, stuck-open transistor P1 disconnects the only normal-1 path to the node while the weak-1 path is maintained. This establishes a weak logic 1 at node A which partially turns on transistors P2 and N2.

3.2.2. Propagation of Weak-0(1) Fault Effects. When a weak fault is sensitized, it can be detected by I_{DDQ} measurement only if the fault effect is propagated by a conducting path from V_{DD} to V_{SS} through the complementary devices, partially turned on due to the weak voltage level. Whether or not such a path is established, in general, depends on the logic state of other nodes in the circuit. For example, in Figure 5, a sensitized weak-1 fault at node A is propagated only if I3 = 1.

It should be noted that some weak faults may be undetectable. We have observed that in some circuits a sensitized weak fault cannot be propagated under any input conditions.

4. Hierarchical Methodology

Switch level logic simulation of the full circuit can obviously be used for the identification and selection of I_{DDQ} vectors. The logic state information can be directly used to determine the faults that will be sensitized and/or propagated upon the application of a test vector. Full circuit switch level logic simulation is, however, not generally feasible and the hierarchical methodology obviates its need. This methodology facilitates the deduction of required switch level information while still employing gate level logic simulation. While the present implementation is targeted towards standard cell library based design approaches, the hierarchical methodology can also be used for switch level designs without pre-defined logic component boundaries. This is because switch level simulators also effectively define logic component boundaries by extracting clusters of channel connected transistors before simulation [16].

We first characterize each logic component type (e.g., a NOR gate or a D flip-flop) to relate logic levels at its inputs/outputs (and internal state nodes, if any) to the detection of leakage faults or the sensitization and/or propagation of weak faults. This information is stored in fault table(s) corresponding to each component type (Section 4.1). This step does involve switch level logic simulation, but it is employed for each component type separately and only once.

We then employ full circuit gate level logic simulation and capture only the inputs/outputs (and internal state nodes, if any) of each component instantiation upon the application of each vector. The fault tables generated earlier are then used to infer the detection of faults in the entire circuit upon the application of each vector. The fault detection inferences are then used to select test vectors for I_{DDQ} testing (Section 4.2). Since we assume that leakage faults and weak faults do not change the logic value at any node in the circuit, logic simulation alone is sufficient to infer the detection of these faults.

4.1. Fault Table Generation

The generation of the necessary fault tables for our fault models is given next.

4.1.1. Leakage Faults. Each component type is simulated at the switch level with exhaustive input stimulus and the logic states at the terminals of each transistor are captured. A leakage fault table is then generated by analyzing the captured logic states to record the leakage fault detection information. For example, if the logic values at the terminals of an NMOS transistor are (G,S,D,B) = (1,0,0,0), leakage faults f_{GS}, f_{GD} and f_{BG} will be detected by I_{DDQ} measurement.

The leakage fault table is configured as an $m \times n$ matrix M, where $m = 2^k$ (k is the number of I/O pins of the component), and n is the number of transistors in the component. Each logic state of the component corresponds to a unique row in the matrix. For example, the logic state (I1,I2,O1) = (0,0,0) of a component corresponds to the first row ($i = 0$) and the state (I1,I2,O1) = (1,1,1) corresponds to the eighth row (i=7) in its leakage fault table. Each entry m_{ij} of the matrix is an octal value which contains leakage fault detection information. It corresponds to a six bit binary number where each bit represents to a leakage fault as shown in Figure 6. Each bit value is set to 1(0) if the corresponding leakage fault will be detected (not detected) by measuring I_{DDQ} in the component I/O state corresponding to the table row.

As an example, consider the two input NOR gate shown in Figure 7. The results of switch level logic simulation are given in Figure 8a and the leakage fault table is given in Figure 8b. The entry m_{53} is $(26)_8 = (010110)_2$ and implies that when (I1,I2,O1) = (1,0,0), i.e., i=4, leakage faults f_{BD}, f_{SD} and f_{GD} in transistor P1 of the gate will be detected if I_{DDQ} is measured. This can be verified from the logic simulation results in Figure 8a.

Note that some rows of the leakage fault tables may correspond to illegal component states. For example, the fourth row (i=3) in Figure 8b would seem to correspond

f_{BG}	f_{BD}	f_{BS}	f_{SD}	f_{GD}	f_{GS}

Fig. 6. Representation of transistor leakage faults.

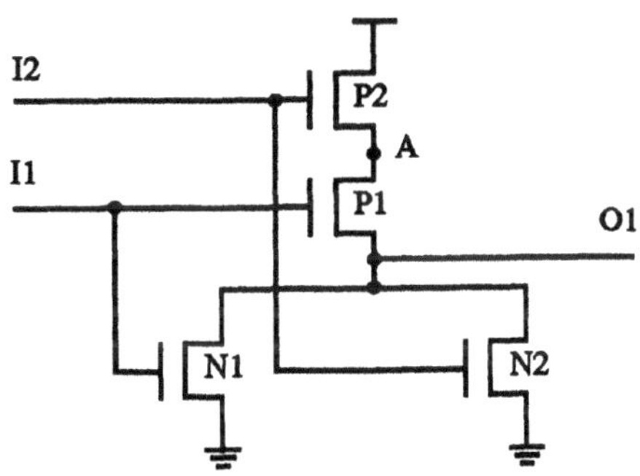

Fig. 7. A logic component for leakage fault characterization.

IIO 121	A
001	1
010	0
100	1
110	Z

(a)

i	N 1	N 2	P 1	P 2
0	0	0	0	0
1	26	26	43	43
2	0	43	70	26
3	0	0	0	0
4	43	0	26	43
5	0	0	0	0
6	43	43	22	0
7	0	0	0	0

(b)

Fig. 8. (a) Logic simulation results for the circuit in Figure 7. (b) Leakage fault table for the circuit in Figure 7.

to a state (I1,I2,O1) = (0,1,1). However, the output O1 of NOR gate would not be logic 1 when the input I2 is logic 1. Therefore, the component would never be in this state. The tables are, however, constructed allocating rows even to illegal states, because such a construction allows indexing through the tables based upon the component I/O state. Obviously, the rows corresponding to the illegal states are never referenced, but the leakage fault table look-up time during test vector selection is reduced.

4.1.2. Weak Faults. A node with weak faults is called an *internal fault node* if there are no conducting paths between the node and any of the outputs of the component. A node with weak faults is called a *boundary fault node* if there is at least one conducting path between the node and one of the component outputs and the node does not drive any transistor gate within the component. A node with weak faults is called an *internal-boundary fault node* if there is at least one conducting path between the node and one of the component outputs and the node also drives at least one transistor gate within the component. For example, in Figure 2, node A is an internal fault node, node O2 is a boundary fault node and node O1 is an internal-boundary fault node.

The sensitization and propagation of a weak fault associated with an internal fault node depends only on the I/O states of the component with the weak fault.

The sensitization of a weak fault associated with a boundary fault node depends on the I/O states of the component with the weak fault. However, the propagation of the fault effect depends on the I/O states of components which are driven by that boundary node.

The sensitization of a weak fault associated with an internal-boundary fault node depends on the I/O states of the component with the weak fault. The propagation of the fault effect may depend on either the I/O states of the component with the weak fault or the I/O states of components which are driven by that internal-boundary fault node.

Up to three tables are generated as the result of characterization of each component type. One table called the *weak fault detection table* contains all detected weak faults associated with internal or internal-boundary fault nodes for each possible I/O state of the component. Another table called the *weak fault sensitization table* contains all sensitized weak faults associated with boundary nodes and sensitized but not detected weak faults associated with internal-boundary nodes for each possible I/O state of the component. The third table is called the *weak fault propagation table*. It shows the input conditions for which a sensitized weak fault at the input of the component will be propagated.

For the component in Figure 9, node A is an internal fault node which can have a weak-0 or a weak-1 fault. The exhaustive stimulus for the characterization of this component and the results of logic simulation are given in Figure 10a. The *weak fault detection* table is given in Figure 10b in which W0(W1) refers to the weak-0(1) fault at node A.

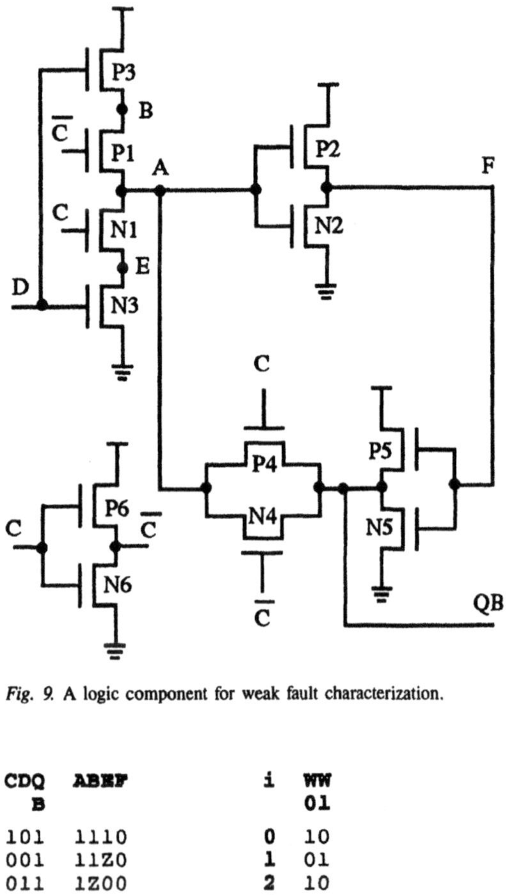

Fig. 9. A logic component for weak fault characterization.

CDQB	ABEF
101	1110
001	11Z0
011	1Z00
110	0001
010	0Z01
000	01Z1

(a)

i	WW 01
0	10
1	01
2	10
3	01
4	00
5	00
6	00
7	00

(b)

Fig. 10. (a) Logic simulation results for the circuit in Figure 9. (b) Weak fault table for the circuit in Figure 9.

4.2. I_{DDQ} *Test Vector Selection*

The full circuit is simulated at gate level using the production test patterns. The logic levels at the inputs/outputs of all components are captured. The I_{DDQ} test vector selection is done using the component fault tables for leakage faults and weak faults. If a vector catches even one new fault, it is selected for I_{DDQ} measurement. It may be noted that for inferring the detection of a weak fault in the circuit, more than one fault table may be referred to.

As an example, the circuit in Figure 11 is simulated at gate level and the captured logic states are given in Figure 12. For the D flip-flop, the leakage fault table is given in Figure 13 and the weak fault detection table is given in Figure 10b. For the NOR gate, the leakage fault table is given in Figure 8b. Since the NOR gate cannot have any weak faults, weak fault detection or weak fault sensitization tables do not exist. Similarly, a weak fault propagation table is not needed for the NOR gate in this circuit as the D flip-flop cannot have a weak fault at QB and I1 is a primary input. By referring to the fault tables, it can be verified that the shaded vectors (Figure 12) cover all leakage faults and weak faults in the circuit. They form the selected set of vectors for I_{DDQ} measurement.

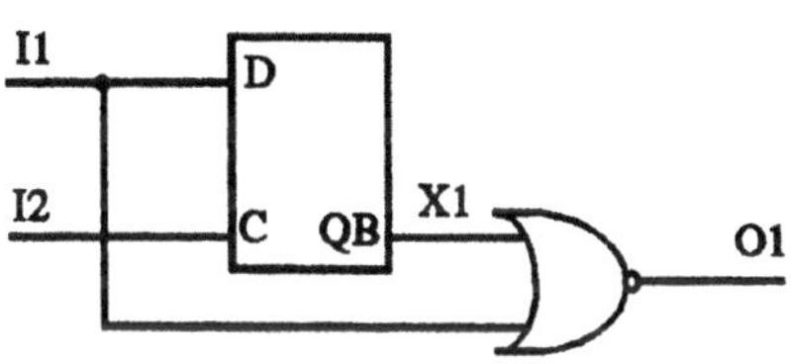

Fig. 11. An example circuit for I_{DDQ} test vector selection.

Time	I1	I2	X1	O1
99	0	1	1	0
199	0	0	1	0
299	1	0	1	0
399	1	1	0	0
499	0	1	1	0
599	1	1	0	0
699	1	0	0	0
799	0	0	0	1
899	0	1	1	0
999	1	1	0	0
1099	1	0	0	0
1199	0	0	0	1
1299	1	0	0	0
1399	1	1	0	0

Fig. 12. Captured logic simulation results for the circuit in Figure 11.

i	N1	N2	N3	N4	N5	N6	P1	P2	P3	P4	P5	P6
0	0	26	0	43	43	26	26	43	43	70	26	43
1	22	43	0	70	26	26	0	26	43	43	43	43
2	0	26	43	43	43	26	22	43	0	70	26	43
3	26	43	43	70	26	26	0	26	0	43	43	43
4	0	0	0	0	0	0	0	0	0	0	0	0
5	70	43	26	33	26	43	43	26	43	0	43	26
6	43	26	43	0	43	43	70	43	26	33	26	26
7	0	0	0	0	0	0	0	0	0	0	0	0

Fig. 13. Leakage fault table for the circuit in Figure 9.

4.3. *A Prerequisite for Weak Faults*

In the previous sections, we have shown that the identification and sensitization of weak faults associated with a node depends upon the normal and weak conducting paths which are connected to the node. In general, it is possible for a node (which is interconnect between two components) to have a weak fault due to a normal conducting path in one component and a weak conducting path in the other component. This can make the characterization of individual component types incomplete and can invalidate the hierarchical analysis methodology for weak faults because the fault can be identified only when the two components are analyzed together. For example, in the circuit of Figure 14 node A can have a weak-1 fault. However, the weak-1 conducting path formed by transistors P1 and N2 is in component 1, while the normal-1 conducting path formed by transistor P2 is in component 2.

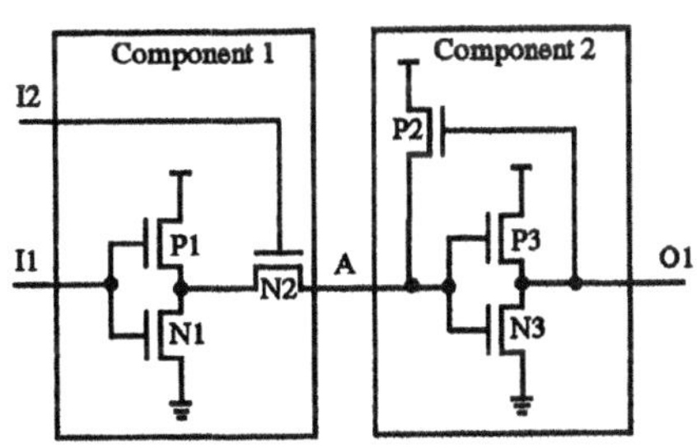

Fig. 14. Normal and weak conducting paths in interfaced logic components.

This clearly indicates the need for guidelines for partitioning the circuit into individual logic components so that the hierarchical methodology can be successfully used. The basic requirement is that all normal and weak conducting paths associated with a weak fault must be contained in a single component. The *sufficient* conditions to guarantee this are:

Table 3. QUIETEST results.

Circuit	Number of Transistors	Number of Leakage Faults	Percentage of Selected Test Vectors	Leakage Fault Coverage	Number of Weak Faults	Percentage of Selected Test Vectors	Weak Fault Coverage
1	7584	39295	0.50%	94.84%	1923	0.35%	85.30%
2	42373	220571	0.99%	90.50%	1497	0.21%	87.64%

(1) all inputs of a component type must only feed the gates of transistors in the component;
(2) the output of different components should not simultaneously drive the same node [14].

It is easy to see that if a CMOS transmission gate is a separate component, it will not meet the above requirement. For example, in Figure 1, input I1 does not directly feed into transistor gates. In this case, we recommend post processing of the circuit netlist where the transmission gate is coalesced into the component which drives it. If the combined component is then characterized, we would avoid the difficulty. These considerations are similar to those in switch-level simulations [16]. In this article, we assume that the above sufficient conditions are satisfied by the logic components in the circuit under test.

5. QUIETEST Implementation and Results

The QUIETEST software system has been implemented in C programming language on a SUN workstation. It supports automated fault table generation and I_{DDQ} test pattern selection for leakage faults and weak faults. The system also includes a fault diagnosis processor for leakage faults. It generates a listing of possible leakage faults based upon the input of failing I_{DDQ} test vectors during testing of a particular part. The user can control the process of vector selection by specifying simulation time intervals during which no vectors should be selected as slowing down of testing for I_{DDQ} measurement may be undesirable. Such control may be necessary to accommodate some dynamic CMOS circuits as they are not compatible with I_{DDQ} testing. QUIETEST can also support CMOS parts in which some nodes may be pulled high (low). It will not select any of the vectors which result in a logic 0(1) at a pulled high (low) node.

QUIETEST was used to select I_{DDQ} vectors for covering leakage faults and weak faults in two VLSI chips. The results are shown in Table 3. In each case QUIETEST selected less than 1% of the total test vectors to provide the same fault coverage as could be obtained by I_{DDQ} testing on 100% of the production test vectors. For circuit 1, 80.2% of the I_{DDQ} vectors selected for testing weak faults were common with the vectors selected for testing leakage faults. The percentage of common I_{DDQ} vectors was even higher (95%) for circuit 2.

6. Conclusion

In this article, we propose an I_{DDQ} test vector selection methodology based upon leakage and weak fault models. The leakage fault model represents defects such as gate oxide shorts and the weak fault model represents defects such as certain opens in CMOS circuits. The conditions for detecting these faults with I_{DDQ} measurements were developed. However, I_{DDQ} measurement upon the application of every test vector is not desirable, as I_{DDQ} measurements at the required level of sensitivity are very time consuming.

Our hierarchical fault analysis methodology makes I_{DDQ} testing feasible in large volume production test environment. The software system QUIETEST selects a small subset of the production test vectors for which I_{DDQ} should be measured. We have used the QUIETEST system to select I_{DDQ} test vectors for two VLSI chips and the results are very encouraging. For both circuits QUIETEST selected less than 1% of the vectors to cover as many leakage and weak faults as would be covered by I_{DDQ} measurement upon the application of every vector. A large percentage of selected I_{DDQ} vectors for testing weak faults were found to be common with the I_{DDQ} vectors selected for testing leakage faults. All test vectors must still be applied for normal production testing. What QUIETEST ensures is that testing need to be slowed down for I_{DDQ} measurements only for a few of these vectors. It should be noted, however, that the maximum attainable leakage or weak fault coverage is limited by the quality of the provided test set.

References

1. S. McEuen, "I_{DDQ} benefits," *Proc. VLSI Test Symp.*, pp. 285-291, 1991.

2. C.F. Hawkins and J.M. Soden, "Electrical characteristics and testing considerations for gate oxide shorts in CMOS ICs," *Proc. Int. Test Conf.*, pp. 544–555, 1985.
3. J.M. Soden, R.K. Treece, M.R. Taylor and C.F. Hawkins, "CMOS IC stuck-open fault electrical effects and design considerations," *Proc. Int. Test Conf.*, pp. 423–430, 1989.
4. J.M. Soden and C.F. Hawkins, "Test considerations for gate oxide shorts in CMOS ICs," *IEEE Design & Test of Comp.*, vol. 3, pp. 56–64, August 1986.
5. C.F. Hawkins and J.M. Soden, "Reliability and electrical properties of gate oxide shorts in CMOS ICs," *Proc. Int. Test Conf.*, pp. 443–451, 1986.
6. J.M. Soden and C.F. Hawkins, "Electrical properties and detection methods for CMOS IC defects," *Proc. 1st European Test Conf.*, pp. 159–167, 1989.
7. F.J. Ferguson, M. Taylor and T. Larrabee, "Testing for parametric faults in static CMOS circuits," *Proc. Int. Test Conf.*, pp. 436–443, 1990.
8. Y.K. Malaiya and S.Y.H. Su, "A new fault model and testing technique for CMOS devices," *Proc. Int. Test Conf.*, pp. 25–34, 1982.
9. W. Maly and P. Nigh, "Built-in current testing: Feasibility study," *Proc. Int. Conf. on Computer-Aided Design*, pp. 340–343, 1988.
10. L.K. Horning, J.M. Soden, R.R. Fritzemeier and C.F. Hawkins, "Measurements of quiescent power supply current for CMOS ICs in production testing," *Proc. Int. Test Conf.*, pp. 300–308, 1987.
11. C. Crapuchettes, "Testing CMOS I_{DD} on large devices," *Proc. Int. Test Conf.*, pp. 310–315, 1987.
12. M. Keating and D. Meyer, "A new approach to dynamic I_{DD} testing," *Proc. Int. Test Conf.*, pp. 316–321, 1987.
13. W. Mao, R.K. Gulati, D.K. Goel and M.D. Ciletti, "QUIETEST: A quiescent current testing methodology for detecting leakage faults," *Proc. Int. Conf. on Computer-Aided Design*, pp. 280–283, 1990.
14. R.K. Gulati, W. Mao and D.K. Goel, "Detection of *undetectable* faults using I_{DDQ} testing," *Proc. Int. Test Conf.*, pp. 770–777, 1992.
15. N. Weste and K. Eshraghian, "Principles of CMOS VLSI design," Reading, MA: Addison-Wesley, 1985.
16. Special issue on "Switch level techniques," *IEEE Design and Test of Comp.*, vol. 4, August 1987.

Weiwei Mao received the B.S. and M.S. degrees in Electrical Engineering from Fudan University, Shanghai, People's Republic of China, in 1982 and 1984, and Ph.D. in Electrical Engineering from the University of Colorado, Colorado Springs, in 1991.

He is presently working as a Principal Member in the CAE & Advanced Technology organization at Ford Microelectronics, Inc. in Colorado Springs. From 1986 to 1990, he was a Research Assistant in the Electrical and Computer Engineering department, University of Colorado at Colorado Springs. From 1984 to 1986, he was an Assistant Professor in the Electronic Engineering department at Fudan University.

Dr. Mao's research interests include I_{DDQ} testing, delay faults testing, automatic test generation, design for testability and testing of mixed signal circuits. He has widely published in the area of testing.

Ravi K. Gulati obtained his B.E. degree in Electrical and Electronics Engg. from B.I.T.S., Pilani, India in 1980. He worked as a Design Engineer at R&D Labs, Tata Electric Cos., Bombay from 1980 to 1983. He obtained the M.S. degree in Electrical and Computer Engg. from the University of Iowa, Iowa City in 1985. Since then Mr. Gulati is employed with Ford Microelectronics, Inc., at Colorado Springs, where presently he is Engineering Manager in the CAE & Advanced Technology organization.

Mr. Gulati has published and taught numerous tutorials on various aspects of testing including I_{DDQ}. He has served on the program committee of the 1991 IEEE VLSI Test Symposium. Presently, he is a program committee member of the International Test Conference and coordinates the area of CAE for defect detection and I_{DDQ} testing. Mr. Gulati is a member of the IEEE and its P1149.4 working group for the development of mixed-signal test bus standards.

JOURNAL OF ELECTRONIC TESTING: Theory and Applications, 3, 359-366 (1992)

Generation and Evaluation of Current and Logic Tests for Switch-Level Sequential Circuits*

CHUN-HUNG CHEN AND JACOB A. ABRAHAM
Computer Engineering Research Center, The University of Texas at Austin, 2201 Donley Dr., Ste. 395, Austin, TX 78758

Received February 7, 1992. Revised August 11, 1992.
Editor: R.K. Gulati

Abstract. This article presents an approach to developing high quality tests for switch-level circuits using both current and logic test generation algorithms. Faults that are aborted or undetectable by logic tests may be detected by current tests, or vice versa. An efficient switch level test generation algorithm for generating current and logic tests is introduced. Clear definitions for analyzing the effectiveness of the joint test generation approach are derived. Experimental results are presented for demonstrating high coverage of stuck-at, stuck-on, and stuck-open faults for switch level circuits when both current and logic tests are used.

Keywords: Current tests, I_{DDQ}, logic tests, test generation.

1. Introduction

VLSI circuits using CMOS technology have been increasing in complexity and comprise a large portion of chips being manufactured today. In order to detect manufacturing defects and achieve high quality in shipped products, tests need to achieve high fault coverage for shorts and opens [1]-[3]. Conventional logic test generation using the stuck-at fault model is not able to produce high quality tests since the tests may not detect all defects which cause shorts and opens. Logic test generation algorithms for switch-level circuits in CMOS technology have been under development for many years. Jain and Agrawal [4] used an equivalent gate model for deriving tests for combinational CMOS circuits. Reddy et al. [5] used Bryant's [6] switch-level model and presented a test generation algorithm for detection of transistor stuck-on and stuck-open faults for combinational MOS circuits. Cho and Bryant [7] proposed derived boolean functions for test generation on sequential MOS circuits. These functions represent the behavior of the good and faulty circuits over a sequence of symbolic input patterns. Chen and Abraham [8] developed a nine-valued relaxation algorithm to generate tests for stuck-at, stuck-on and stuck-open faults for mixed switch and gate level sequential circuits.

Recent work, however, has shown that logic-level tests may not be able to assure the detection of all CMOS integrated circuit (IC) physical defects [10], in particular those which cause high quiescent current. Even if the defective chip always produces the correct logic value, the high current may indicate reliability problems and potential for future logic failures. Recent research [11]-[16] shows that the detection of many CMOS manufacturing defects can be improved by monitoring the quiescent power supply current (I_{DDQ}). Normally, the leakage current of CMOS circuits under a quiescent state is very small but increases in the presence of many defects. I_{DDQ} monitoring together with conventional logic tests, can result in high quality tests for CMOS circuits.

Only limited work has been done towards test generation for I_{DDQ} testing. In [12], the detection of X states is used to derive the test vectors, where such states result from a node that is connected to both power and ground. The test vector is generated by the intersection of patterns which enable paths to power and those which connect the node to ground. However, the effectiveness of the tests generated using this method was not established. An approach for generating I_{DDQ} test sets to cover bridges with supply and ground was proposed in [15]. Since the fault model does not include the internal node faults and stuck-on/stuck-open faults

*This is expanded version of the work originally presented at the 1991 International Test Conference.

for the CMOS circuit components, the generated I_{DDQ} test sets do not provide a complete solution for the detection of CMOS circuit faults. Another approach [16] includes bridging faults and stuck-on faults in the fault model. It uses a random pattern testing approach for I_{DDQ} testing. For some of the ISCAS 85 combinational circuits [17], it takes up to 1/2 million pseudo-random vectors, a very large number of tests for relatively small circuits. Mao et al. [18] proposed a methodology for selecting a small number of test vectors from the full test set for I_{DDQ} testing to detect leakage faults.

In this article, we propose an efficient approach to developing high quality tests for switch-level circuits using both current and logic test generation algorithms. We are interested in the detection of CMOS physical failures such as stuck-at, stuck-on, and stuck-open faults, which contribute to logic and quiescent current errors. Figure 1 shows the need for using both current test generation (CTG) and logic test generation (LTG) algorithms. The stuck-open fault of transistor P3 can be detected by LTG. It may be detected by measuring the transient response of the stuck-open fault [3] using CTG; however, this cannot be guaranteed in all cases. On the other hand, the stuck-on fault of transistor P1 is undetectable using LTG (because of the *X*-value at the output of the gate), while it can be easily detected by CTG.

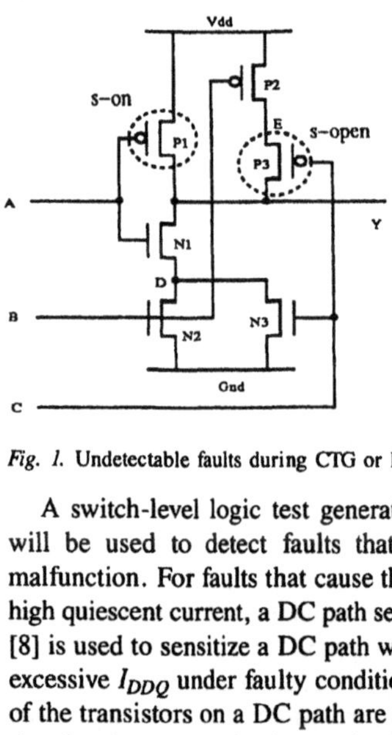

Fig. 1. Undetectable faults during CTG or LTG.

A switch-level logic test generation algorithm [8] will be used to detect faults that can cause logic malfunction. For faults that cause the circuits to draw high quiescent current, a DC path sensitization method [8] is used to sensitize a DC path which will result in excessive I_{DDQ} under faulty conditions. The strengths of the transistors on a DC path are used as an indication for the current in the conducting DC path. The test vectors for current testing are generated using the DC paths which will cause the most increase in I_{DDQ} in the presence of a target fault. To improve the performance of current test generation, concurrent fault simulation for current tests is used to drop other faults which will cause high I_{DDQ}. This reduces the size of test vectors and CPU time significantly.

We generate tests for complex switch-level circuits using both current and logic test generation approaches. Since these two test generation algorithms are different, the aborted or undetectable faults from one test generation approach may turn out to be detectable by the other. Therefore the individual fault coverage and test generation efficiency measures are not adequate for evaluating the combined algorithms. We therefore present new definitions for analyzing the effectiveness of tests when using both approaches. In the following section, we will provide some preliminaries, and will describe current and logic test generation algorithms in more detail in Section 3. Clear definitions for describing the performance of combined test generation approaches are also derived. Section 4 describes results on a set of switch-level circuits which demonstrate the need to use both logic and current tests in order to achieve high quality. Concluding remarks are given in Section 5.

2. Preliminaries

Logic test generation for switch-level sequential circuits is very complex. The effect of a fault has to be propagated to a primary output possibly using several time frames in order for it to be observed. When the fault propagation is successful, we then have to go through the justification process to find consistent input values. Test generation for I_{DDQ} testing is easier than logic test generation since it is not necessary to perform the fault propagation during test generation. Only the justification process is required. The observation of a fault effect is achieved by monitoring the excessive power supply current. Redundancies (in a logic sense) which cause faults to be undetectable during conventional logic test generation often cause no problem during I_{DDQ} test generation. In the following, we give some necessary terminology, with particular emphasis on switch-level circuits and current test generation.

DC coupling components (DCCs), the set of transistors that have drain-source connections between any two nodes, are constructed on top of a CMOS circuit. Each DCC can have a pull-up, a pull-down and a pass transistor group. The center node of a DCC block is the node from where these groups originate. Figure 2 shows a DCC which contains these groups and node *D* is a center node. A *pull-up group* consists of pMOS

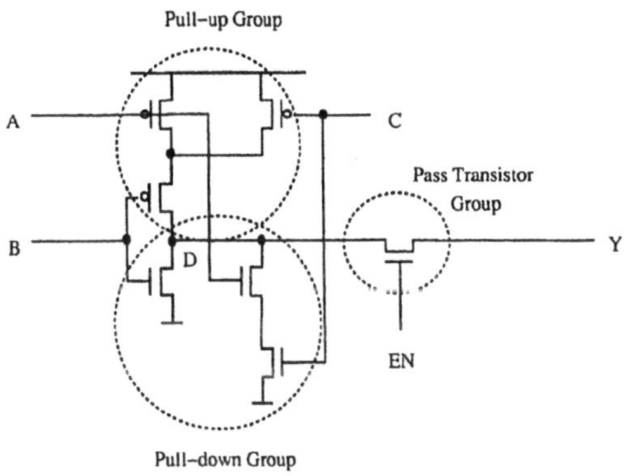

Fig. 2. Example of a DCC component.

transistors which are part of DC paths connecting the center node to power. A *pull-down group* consists of nMOS transistors which are part of DC paths connecting the center node to ground. The conduction and strength of a DC path can be determined by the logic values at the transistor gates and the conductance of the transistors in the pull-up and pull-down groups.

For a static CMOS circuit, the quiescent power supply current is very small. Various defects such as shorts, opens, and stuck-at faults can cause an increase in I_{DDQ}. In I_{DDQ} testing, a fault α is said to be *detected* if a test causes the faulty circuit to have a DC path between power and ground in the presence of the fault, and it draws an excessive I_{DDQ} along the DC path. A fault β is said to be *undetectable* if it causes no excessive power supply current under all possible input sequences. A fault γ is said to be *aborted* if the specified time limit is exceeded before identifying this fault as detected or undetectable. A fault is *logic redundant* if it does not produce a logic malfunction under all possible input sequences, and therefore is undetectable by conventional logic tests. Similarly, a fault is *I_{DDQ} undetectable* if it will never cause excessive I_{DDQ} in the circuit, and therefore is undetectable by current tests. Test generation *efficiency* is computed as the sum of the number of detected faults and the number of faults proven to be undetectable, divided by total number of faults.

A stuck-at-0 fault on the gate of a pMOS transistor is equivalent to a stuck-on transistor fault. Likewise, a stuck-at-1 fault on the gate of a nMOS transistor is equivalent to a stuck-on transistor fault. They may be detected by current or logic tests. Stuck-open faults in CMOS transistors can cause timing faults in the circuit when they act as weakly-on active loads [11]. Such faults may not be detected by current tests, but two-pattern logic tests may be derived for them.

3. Test Generation

We have observed that testing complex CMOS circuits using logic tests alone may not be sufficient and it is advantageous to combine I_{DDQ} testing with logic testing. Faults that cause logic errors on a primary output and those which cause high I_{DDQ} have to be detected using different approaches. Some of the faults which can cause both incorrect logic values and high I_{DDQ} may be aborted during logic test generation (because of CPU time limits, etc.), but current tests may be found for them. Other faults may not be detectable by monitoring the current, but logic tests may be generated for them. Therefore, in order to get high quality tests, it is advantageous to adopt both current and logic test generation approaches for testing CMOS circuits.

The sets of faults that are detected, undetectable, and aborted during test generation using the two different approaches may be different or may overlap. Thus, the fault coverage and test generation efficiency measures which are adequate for these test generation approaches individually may not be appropriate in a combined approach. Evaluation of the effectiveness of the joint test generation approach is discussed in the following.

3.1. Evaluation of Joint Test Generation Effectiveness

A joint test generation approach requires a modification of conventional measures of fault coverage and test generation efficiency. We begin with characterizing the various possibilities during both current and logic test generation. Figure 3 shows a fault status diagram (FSD) for various results of tests (such as faults being detected, undetectable or aborted) for current test generation (CTG) and logic test generation (LTG). When a circuit is tested using patterns from both CTG and LTG approaches, it is said to be *good* if it passes both tests. We define A1 to A9 in Figure 3 for evaluating joint test generation effectiveness.

Faults within A3 to A7 will be detected by either I_{DDQ} tests or logic tests. Therefore, if F_{Ai} is the number of faults in fault set A_i, then

$$\text{Joint Fault Coverage} = \frac{F_{A3} + F_{A4} + A_{A5} + F_{A6} + F_{A7}}{\textit{Total Faults}}$$

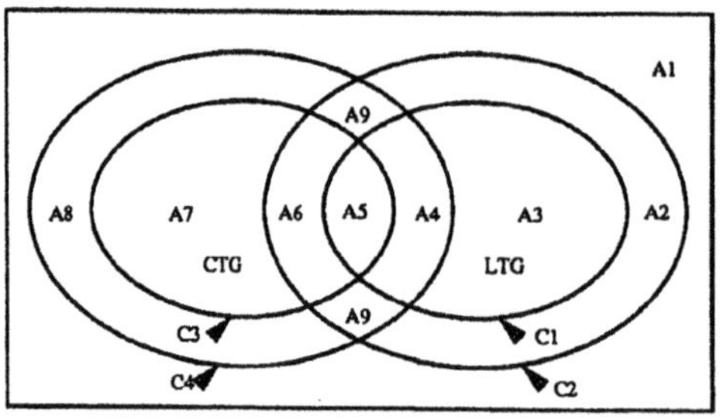

C1 : faults within C1 are the set of faults detected by LTG
C2 : faults between C1 and C2 are the set of faults aborted by LTG
C3 : faults within C3 are the set of faults detected by CTG
C4 : faults between C3 and C4 are the set of faults aborted by CTG
A1 : faults that are undetectable by current tests and logic tests
A2 : faults that are aborted during LTG and undetectable by current tests
A3 : faults that are detected by logic tests alone
A4 : faults that are aborted during CTG and detected by logic tests
A5 : faults that are detected by current tests and logic tests
A6 : faults that are aborted during LTG and detected by current tests
A7 : faults that are detected by current tests alone
A8 : faults that are aborted during CTG and undetectable by logic tests
A9 : faults that are aborted during CTG and LTG

Fig. 3. Fault status diagram.

Likewise, the joint test generation efficiency is defined as the sum of the detected faults (by I_{DDQ} tests or logic tests) and faults proven to be undetectable by both tests, divided by the total number of faults. Alternatively, in terms of the aborted faults,

$$\text{Joint Test Generation Efficiency} = 1 - \frac{F_{A2} + F_{A8} + F_{A9}}{\textit{Total Faults}}$$

3.2. Test Generation for Logic Faults

A nine-valued relaxation algorithm [8] is used for generating logic tests for stuck-at, stuck-on, and stuck-open faults for CMOS circuits. This algorithm uses the nine-valued algebra and strength operations together with DC path sensitization during the relaxation process.

Before the test generation process starts, the direction of the signal flow for pass transistors is assigned based upon the circuit topology. Those for which flow directions cannot be assigned are the truly bidirectional pass transistors. Appropriate flow directions are assigned to these pass transistors during the test generation process. These assignments are made either during the sensitization process or the justification process, but not both. The bidirectional pins are assigned as primary inputs or primary outputs, as necessary for test generation, for each time frame, since most testers have the capability to drive or observe a pin as desired during each clock.

All the nodes are set to unknown at the beginning of test generation. Multiple paths are sensitized from a fault to an output, and the extended back trace concept [9] is used to justify values and to reduce the complexity of the test generation process. Some of the faults (such as stuck-open faults) require two-pattern tests to be detected, and such sequences are automatically found by the algorithm. The presence of a fault is also considered in the initialization and justification process to guarantee that the tests generated are valid.

3.3. Test Generation for I_{DDQ} Faults

We use a DC path sensitization method to derive the initial justification patterns for a transistor fault. To sensitize a fault, both DC path blocking and gate value sensitization are used to achieve a conducting path from power to ground. The justification of the line values during CTG is similar to logic test generation justification process.

As in conventional test generation, fault simulation for I_{DDQ} testing is used to identify the faults that will be detected using the same I_{DDQ} test. The conventional concurrent fault simulation concept is extended to find all detectable I_{DDQ} faults during a single pass fault simulation. Using this approach, the number of test sequences and the test generation CPU time is significantly reduced.

- *Strength operations*. The strength of a transistor can be defined as inversely proportional to its source to drain resistance when it is on. The logic value of a DCC is derived from the logic states and the strengths of the transistors in the pull-up group and pull-down group. The strength of a DC path is then used as an indication of the I_{DDQ} magnitude.

  A DCC can be abstracted into two switches as shown in Figure 4. The output logic value of the DCC can be obtained by the states and strengths of the switches.
- *DC path sensitization for* I_{DDQ} *testing*. When the initial justification patterns for a fault in a DCC block are derived, all the DC paths in the DCC block are ordered according to the strength of each path so that the one which will result in the maximum I_{DDQ} will be selected first.

  For a transistor stuck-on fault in the pull-up group of a DCC block, a DC path that includes the faulty

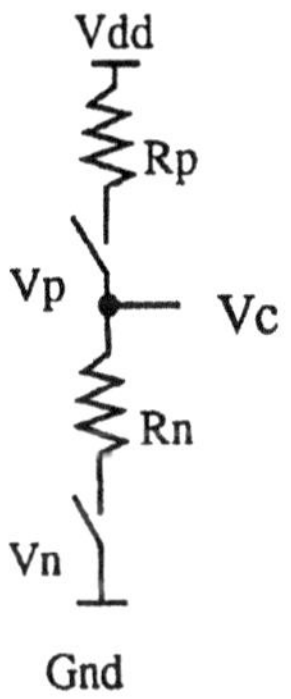

Vp	Vn	Vc		
		Rp>Rn	Rp=Rn	Rp<Rn
0	0	X	X	X
0	1	0	0	0
0	X	X	X	X
1	0	1	1	1
1	1	0	X	1
1	X	X	X	1
X	0	X	X	X
X	1	0	X	X
X	X	X	X	X

Fig. 4. Three-valued logic operation with different strengths of DC paths.

transistor is selected. The gate value of the faulty transistor is set to logic value one and all the other paths in the pull-up group are blocked. A path in the pull-down group is then selected for completing an I_{DDQ} path from power to ground.

For an internal node stuck-at-1 fault in the pull-up group of the DCC block, a DC path that includes the faulty node is selected. All the other paths in the pull-up group connected to power are blocked. A path in the pull-down group is then selected for completing a DC path from the faulty node to ground. For an internal node stuck-at-0 fault in the pull-up group of the DCC block, all the paths in the pull-down group connected to ground are blocked. A path in the pull-up group is selected for completing a DC path from power to the faulty node.

The transistor faults that affect the pull-down group of a DCC block can be detected in a similar fashion.

For the stuck-at-1(0) faults at the output of a DCC block, the DC paths in the pull-up group (pull-down group) are all blocked and a path in the pull-down group (pull-up group) is selected for sensitizing a DC path from the faulty output to ground (power).

- *Fault simulation for* I_{DDQ} *testing.* We use a similar approach as in conventional concurrent fault simulation to implement fault simulation for I_{DDQ} testing. Each DCC block is associated with a list of faulty machines. Each faulty machine represents a single fault effect with respect to a DCC block. These machines are preprocessed so that each faulty machine has sensitized input patterns derived using DC path sensitization. When fault simulation starts, a faulty machine will be activated if the input pattern of the DCC block matches the sensitized input pattern of the faulty machine. This process continues until there are no more events for current fault simulation.
- *Test generation approach.* Before test generation, all the possible input patterns for every fault in the list are generated using DC path sensitization. At the beginning of test generation, all the node values are set to unknown. An initial justification pattern of the target fault is selected and the line values to be justified are set. The justification process of test generation is then used to find a justification sequence so that all the output values of the latches can be in the unknown state initially. This process is continued until the justification process succeeds (fault is detected), no more choices are available from the initial justification patterns (fault is undetectable), or the specified time limit is exceeded (fault is aborted).

When the justification process succeeds, if the concurrent I_{DDQ} fault simulation is invoked, other detected I_{DDQ} faults are dropped from the fault list. The test generation continues until there is no fault left in the list.

4. Results

In order to quantitatively compare the two approaches, we use existing switch-level examples [8], including shifters, counters, adders, multipliers and PLA finite-state machines. Table 1 shows the statistics of the example circuits that we use for both current and logic test generation. The number of inputs (no. ips), outputs (no. ops), bi-directional inputs (no. bios), transistors (no. trs), and latches (no. latches) are listed. Sft 8 is a shifter circuit. Ctr32 is a 32-bit counter; this was chosen as a difficult example of a sequential switch-level circuit. The multiplier bmult is a single stage modified Booth multiplier which contains a coefficient latch, recoding logic and a programmable add-subtract capability [19]. Mult8 is an 8-bit multiplier. Adder1 is

Table 1. Statistics for the example circuits.

ex ckts	no. ips	no. ops	no. bios	no. trs	no. latches
sft8	2	8	0	38	16
ctr32	33	32	0	320	32
bmult	12	3	4	67	17
mult8	33	24	4	535	136
adder1	24	9	0	224	0
adder2	4	16	0	128	32
pla	32	24	0	400	56

an 8-bit combinational ripple adder (all the other circuits are sequential). Adder2 is an 8-bit adder-counter circuit. Pla is a PLA-based finite-state machine. Table 2 shows the breakdown of faults in the example circuits.

Table 2. Breakdown of faults in the example circuits.

ex ckts	s-at flts	s-on flts	s-open flts	total flts
sft8	72	38	38	148
ctr32	450	320	320	1090
bmult	146	67	67	280
mult8	1182	535	535	2252
adder1	258	224	224	706
adder2	170	128	128	426
pla	576	400	400	1376

A mixed-level sequential test generator [8] is used to generate logic tests for stuck-at faults, transistor stuck-on faults, and transistor stuck-open faults in these switch-level circuits. All the nodes were set to unknown at the beginning of test generation. The time frame limit was set to 30, except for Ctr32, for which the time frame was set to 50 due to the high sequential depth. The CPU time limit in all cases was 20 seconds. We assumed that faults in the clock control logic could be detected using the individual initial value propagation method [20], and these faults were not considered. All the test generation results are from a SUN 4/280.

Table 3 shows the performance of logic test generation and current test generation for line stuck-at faults, transistor stuck-on faults, and stuck-open faults for the example circuits. No fault simulation was done during either the current test generation or logic test generation for a fair comparison. Each test circuit has two rows of results. The first row gives the result from logic test generation, and the second row the result from current test generation. The detected faults (det flts) are given in the first column. The undetectable faults (und flts) are those identified by the algorithm as definitely undetectable (after searching all possibilities). Some faults were aborted (abt flts) when the specified time limit was exceeded. As can be seen, current test generation takes less CPU time and requires fewer test patterns because the test generation process is less complex. The results also show that there were no aborted faults during current test generation.

Table 3. Test generation results for LTG and CTG.

ex ckts	det flts	und flts	abt flts	test ptns	CPU time (sec)
sft8	148	0	0	1381	2916 (LTG)
	97	51	0	344	5.38 (CTG)
ctr32	494	20	576	19850	14201.02 (LTG)
	844	246	0	1722	62.48 (CTG)
bmult	108	172	0	459	22.22 (LTG)
	189	91	0	315	7.18 (CTG)
mult8	824	1428	0	3436	751.38 (LTG)
	1380	872	0	2700	176.83 (CTG)
adder1	384	294	28	638	818.18 (LTG)
	472	234	0	472	36.94 (CTG)
adder2	394	32	0	2400	101.55 (LTG)
	320	106	0	410	10.60 (CTG)
pla	1376	0	0	5760	493.57 (LTG)
	1064	312	0	1320	48.81 (CTG)

Table 4. Mutually exclusive fault sets of CTG and LTG.

ex ckts	total flts	A1 flts	A2 flts	A3 flts	A5 flts	A6 flts	A7 flts
sft8	148	0	0	51	97	0	0
ctr32	1090	5	141	134	360	435	15
bmult	280	46	0	45	87	0	102
mult8	2252	538	0	334	556	0	824
adder1	706	82	6	146	286	22	164
adder2	426	8	0	98	304	0	16
pla	1376	0	0	312	1064	0	0

The mutually exclusive fault sets are shown in Table 4. Table 5 compares the effectiveness of using logic tests (l.t.) and current tests (c.t.). The table compares results when (a) logic test generation is done first and current test generation is attempted for the remaining faults, and (b) when current test generation is done first, followed by logic test generation. The second column (flt det) gives the percentage of faults that are detected by LTG. The third column gives the percentage of faults that are aborted by logic tests and detected by current tests. The fourth column gives the percentage of faults that are declared redundant (undetectable) during logic test generation, but are detected by current tests. The fifth column gives the percentage of faults that are detected by current tests. The sixth column gives the

Table 5. Effectiveness of logic and current test (in %).

	logic test gen.			current test gen.	
ex ckts	flt det	abt & det by c.t.	und & det by c.t.	flt det	und & det by l.t.
sft8	100	0	0	65.54	34.46
ctr32	45.32	39.91	1.38	77.43	12.29
bmult	38.57	0	36.43	67.50	16.07
mult8	36.59	0	36.59	61.28	14.83
adder1	54.39	3.11	23.23	66.86	20.68
adder2	92.49	0	3.76	75.12	23.00
pla	100	0	0	77.33	22.67

percentage of faults that are undetectable by current tests but are detected by logic tests. The results show that logic and current test generation are effective for different types of faults, and that use of both approaches is necessary for the detection of all the CMOS circuit faults.

Table 6. Joint test generation results (in %).

ex ckts	joint fcov.	joint eff.
sft8	100	100
ctr32	86.61	87.06
bmult	83.57	100
mult8	76.11	100
adder1	87.54	99.15
adder2	98.12	100
pla	100	100

Table 6 shows the joint fault coverage (joint fcov.) and test generation efficiency (joint eff.). The joint test generation efficiency is very high for all the test circuits except Ctr32. This circuit has 141 faults that may cause logic errors but will not cause excessive I_{DDQ}. Note that Ctr32 has 435 aborted faults (A6) being detected by I_{DDQ} testing. It should be pointed out that the test generator had no information of the structure or behavior of the 32-bit counter, and had to deal with a flat, switch-level netlist and all stuck-at, stuck-on, and stuck-open faults. If more information about the structure and behavior were available, or if the counter were designed for testability (by partitioning it, etc.), the performance of test generation could be improved.

In addition to achieving high fault coverage, another test goal is to reduce the test costs. This includes both test generation time as well as the number of test patterns (which contributes to cost of applying the tests). A trade-off between logic tests and I_{DDQ} tests can be made in this arena. The overall cost of testing a circuit depends on the test pattern generation time, the fault coverage required, tester time for each unit, and the volume of circuits under test. While current test generation takes much less CPU time and fewer test patterns, testing a circuit using I_{DDQ} tests is several orders of magnitude slower than the logic tests. For example, given the required joint fault coverage (as shown in Table 6), there are two extreme cases that we can study. In Table 7, case 1 shows the number of test vectors and the CPU time necessary when tests are generated for the circuit using LTG first and then making use of CTG for the aborted and redundant faults. Case 2 shows the results of using CTG first and then making use of LTG for the redundant faults (there are no aborted faults in the example circuits when using CTG). As can be seen from Figure 3, fault set A5 is where the trade-off takes place. Performing LTG first requires more CPU time, but results in fewer patterns for current tests, reducing overall tester time. On the other hand, performing CTG first reduces the CPU time necessary, but results in more current tests and higher tester time. The overall trade-off, of course, depends on the costs of computer and tester time, and the volume of parts produced.

Table 7. Two extreme cases for joint test generation.

	Case 1 (LTG first)			Case 2 (CTG first)		
ex ckts	LTG ptns	CTG ptns	CPU time	CTG ptns	LTG ptns	CPU time
sft8	1381	0	29.16	344	476	15.4
ctr32	19850	914	14226	1722	5384	1761
bmult	459	170	24.83	315	190	10.7
mult8	3436	1612	816.08	2700	1392	288.3
adder1	638	186	826.76	472	246	206.1
adder2	2400	20	101.95	410	552	33.9
pla	5760	0	493.57	1320	1306	160.7

5. Conclusion

We have presented techniques for obtaining high quality tests for switch-level circuits using both current and logic test generation algorithms. The combination of these approaches provides a better assurance for the detection of physical failures in CMOS circuits than using either one alone. Clear definitions for analyzing the effectiveness of the joint test generation approach allow the evaluation of the performance for the test generation algorithm. Trade-off between logic tests and current tests can be made for optimizing test generation time and production test time. The results from the example circuits show the effectiveness of the joint test generation approach.

Acknowledgment

This research was supported by the Semiconductor Research Corporation under Contract 91-DP-142.

References

1. M.W. Levi, "CMOS is most testable," *Int. Test Conf.*, pp. 217–220, October 1981.

2. M.K. Reddy, S.M. Reddy, "Detecting FET stuck-open faults in CMOS latches and flip-flops," *IEEE Design and Test of Computers*, pp. 17-26, October 1986.
3. J.M. Soden, R.K. Treece, M.R. Taylor and C.F. Hawkins, "CMOS IC stuck-open fault electrical effects and design considerations," *Int. Test Conf.*, pp. 423-430, September 1987.
4. S. Jain, V.D. Agrawal, "Test generation for MOS circuits using D-algorithm," *20th Design Automation Conf.*, pp. 64-70, June 1983.
5. M.K. Reddy, S.M. Reddy and P. Agrawal, "Transistor level test generation for MOS circuits," *22nd Design Automation Conf.*, pp. 825-828, June 1985.
6. R. Bryant, "A switch-level model and simulator for MOS digital systems," *IEEE Trans. on Computer*, vol. C-33, pp. 160-177, February 1984.
7. K. Cho, R. Bryant, "Test pattern generation for sequential MOS circuits by symbolic fault simulation," *26th Design Automation Conf.*, pp. 418-423, June 1989.
8. C.H. Chen and J.A. Abraham, "Mixed-level sequential test generation using a nine-valued relaxation algorithm," *Int. Conf. on Computer-Aided Design*, pp. 230-233, November 1990.
9. R.A. Marlett, "EBT, A comprehensive test generation technique for highly sequential circuits," *17th Design Automation Conference*, pp. 332-339, June 1978.
10. J.M. Soden and C.F. Hawkins, "Electrical properties and detection methods for CMOS IC defects," *European Test Conf.*, pp. 159-167, April 1989.
11. W. Maly, P.K. Nag, and P. Nigh, "Testing oriented analysis of CMOS ICs with opens," *Int. Conf. on Computer-Aided Design*, pp. 344-347, November 1988.
12. P. Nigh and W. Maly, "Test generation for current testing," *European Test Conf.*, pp. 194-200, April 1989.
13. K.J. Lee and M.A. Breuer, "On detecting single and multiple bridging faults in CMOS circuits using the current supply monitoring method," *Int. Symp. on Circuits and Systems*, pp. 5-8, May 1990.
14. J.M. Soden, R.R. Fritzemeier and C.F. Hawkins, "Zero defects or zero stuck-at faults—CMOS IC process improvement with I_{DDQ}," *Int. Test Conf.*, pp. 255-256, September 1990.
15. R.R. Fritzemeier, J.M. Soden, R.K. Treece and C.F. Hawkins, "Increased CMOS IC stuck-at fault coverage with reduced I_{DDQ} test sets," *Int. Test Conf.*, pp. 427-435, September 1990.
16. F.J. Ferguson, M. Taylor, T. Larrabee, "Testing for parametric faults in static CMOS," *Int. Test Conf.*, pp. 436-443, September 1990.
17. F. Brglez and H. Fujiwara, "A neutral netlist of 10 combinational benchmark circuits and a target translator in fortran," *Int. Symposium on Circuits and Systems*, June 1985.
18. W. Mao, R.K. Gulati, D.K. Goel and M.D. Ciletti, "QUIETEST: A quiescent current testing methodology for detecting leakage faults," *Int. Conf. on Computer-Aided Design*, pp. 280-283, November 1990.
19. P. Denyer and D. Renshaw, *VLSI Signal Processing: A bit-serial approach*, Reading, MA: Addison-Wesley, 1985.
20. T. Ogihara, S. Saruyama and S. Murai, "Test generation for sequential circuits using individual initial value propagation," *25th Design Automation Conf.*, pp. 424-427, June 1988.

Chun-Hung Chen received the B.S. degree in Electrical Engineering from National Taiwan University, Taipei, Taiwan in 1976, and received the M.S. degree in Electrical and Computer Engineering from U.C.S.B., Santa Barbara, California, in 1983. From 1983 to 1989 he was engaged in several research projects on simulation, timing verification and synthesis for testability. He has been a research assistant since 1990 at the Computer Engineering Research Center. He is currently working towards the Ph.D. degree at the University of Texas at Austin.

His current research interests include logic simulation, timing verification, design for testability and automatic test pattern generation.

Jacob A. Abraham is a Professor in the Department of Electrical and Computer Engineering at the University of Texas at Austin. He is also director of the Computer Engineering Research Center and holds a Cockrell Family Regents Chair in Engineering. He received the Bachelor's degree in Electrical Engineering from the University of Kerala, India, in 1970. His M.S. degree, in Electrical Engineering and Ph.D., in Electrical Engineering and Computer Science, were received from Stanford University, Stanford, California, in 1971 and 1974, respectively. From 1975 to 1988 he was on the faculty of the University of Illinois, Urbana, Illinois.

Professor Abraham's research interests include fault-tolerant computing, VLSI design and test, computer-aided design, and computer architecture. He is the principal investigator of several contracts and grants in these areas and a consultant to industry and government in the areas of testing and fault-tolerant computing. He has published over 150 papers in journals and conferences, and has supervised 30 Ph.D. dissertations. He was elected Fellow of the IEEE in 1985, and is also a member of ACM and Sigma Xi.

JOURNAL OF ELECTRONIC TESTING: Theory and Applications, 3, 367–375 (1992)

Diagnosis of Leakage Faults with I_{DDQ}

ROBERT C. AITKEN
Design Technology Center, Hewlett-Packard Co., 5301 Stevens Creek Blvd., MS 53U/36, Santa Clara, CA 95052-8059

Received February 7, 1992. Revised August 4, 1992.
Editor: R.K. Gulati

Abstract. Recently there has been renewed interest in fault detection in static CMOS circuits through I_{DDQ} monitoring. This work shows that, in addition to fault detection, accurate fault diagnosis may be performed using a combination of current and voltage observations. The proposed system combines a simple single fault model for test generation with a more realistic multiple defect model for diagnosis, and as a result requires only minor modifications to existing stuck-at fault ATPG software. The associated hardware is sufficiently simple that on-board implementation is possible. Experimental results demonstrate the effectiveness of the method on a standard-cell ASIC.

Keywords: Fault diagnosis, I_{DDQ} testing, leakage fault model.

1. Introduction

The most immediate problem in production testing is fault detection in order to identify the chips with manufacturing defects. Recently, monitoring of the quiescent power supply current (I_{DDQ}) has reemerged as a useful fault detection tool in static CMOS circuits [1], [2], [3], [4], [5]. A few results have also been presented for fault diagnosis—determining which fault(s) is(are) present—(also known as fault location) [6], [7], [8], [9]. Current testing alone has been believed to obscure diagnosis information, since only one observation point is available [10]. However, this article shows that when coupled with standard voltage testing (observing logical values of circuit outputs), I_{DDQ} measurements can be part of an effective diagnosis strategy.

This article introduces a method for performing fault diagnosis in full scan or purely combinational static CMOS circuits using a combination of current and voltage information. Novel features of the method, christened DIVA (Diagnosis using I_{DDQ} and Voltage Analysis), include the combined use of a simple fault model for test set generation and more realistic defect models for diagnosis, plus simple hardware to collect the required data. It is assumed that I_{DDQ} testing will be performed on the circuit under test; this article addresses the question of using its results for diagnosis.

This article is organized as follows: Section 2 discusses the defect models used by DIVA and reviews previously published techniques for diagnosing faults. Section 3 outlines the modifications which need to be made to existing test software to facilitate defect diagnosis with I_{DDQ} testing. DIVA's defect diagnosis algorithm is given in Section 4. Section 5 describes test circuitry which will produce the data required for this algorithm. Section 6 examines some results of experiments designed to test the effectiveness of the DIVA method. Finally, Section 7 concludes the article.

2. Theory and Background

This article assumes circuits which are implemented in static CMOS and are either purely combinational or full scan. It is further assumed that there is no bus contention inherent in fault-free operation, and that the circuits may be modeled at gate level.

2.1. Defect Models

A wide variety of defects in static CMOS will manifest themselves through increased quiescent current. Defects may be characterized as either global (affecting a large part of the chip) or local (affecting at most a few transistors). Most local defects result from contamination during circuit fabrication, which results in either extra or missing material at a given site [11]. Examples of these include gate oxide shorts, unintended bridges between nodes, and missing connections. Such defects, particularly shorts, can result in unintended connections between power and ground. As a result, the circuit draws current in its static state, and the defects can be detected by monitoring the quiescent current (I_{DDQ}). Many defects which may be observed via I_{DDQ} are undetectable with conventional voltage testing [4]. Detection of such defects may improve overall circuit quality levels [12].

The DIVA method described in this article is intended to diagnose defects which cause an immediate increase in I_{DDQ} consumption. The following transistor leakage model is used, although the DIVA methodology is independent of the underlying model—line bridging has also been used successfully. Consider the short shown in Figure 1. Whenever input A is driven to 1, a direct V_{DD} to GND coupling will exist and the quiescent current will increase above its background level. Note that it is not necessary to propagate any signal from Q to a primary output to observe this defect.

Tables may be generated, for diagnostic purposes, to show the effect of leakage between the source (s), drain (d), gate (g), and substrate or bulk (b) of individual transistors in standard CMOS gates. This model is equivalent to that used in [3]. Examples for the inverter

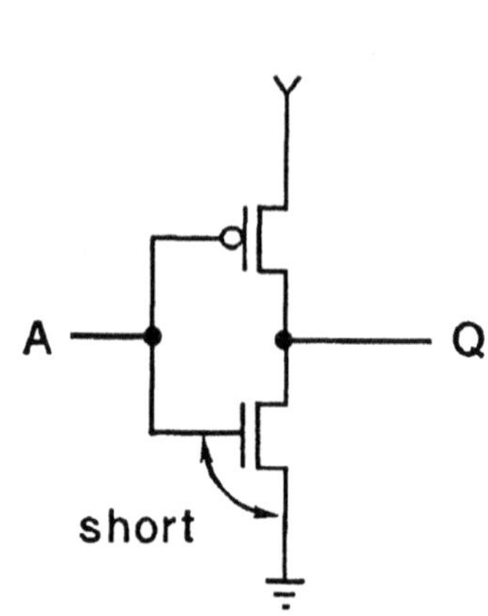

Fig. 1. Leakage fault in inverter.

of Figure 1 and the NAND gate of Figure 2 are given in Table 1. I_{DDQ} is denoted by I, with high current labeled H and background current B. The gate output value is Q and an unknown or indeterminate logic state is indicated by X. An entry of "." indicates no change between faulty and fault-free. Note that the labeling of gate inputs is significant. The column under each modeled fault indicates its I and Q values for the corresponding input values. The defects are grouped into defect classes, which have identical responses for all input values and are thus indistinguishable from a diagnostic perspective.

The following example illustrates the entries: The short shown in Figure 1 is a gate to source (gs) leakage in the N transistor for the inverter in Table 1a. The

Table 1. Modeled behavior of transistor leakages.

a) Inverter

INV			*N* transistor						*P* transistor					
A	I	Q	sd	gs	gd	sb	gb	db	sd	gs	gd	sb	gb	db
0	B	1	HX	..	HX	..	..	HX	..	HX	HX	..	HX	..
1	B	0	..	HX	HX	..	HX	..	HX	..	HX	..	..	HX
defect class			1	2	3	-	2	1	2	1	3	-	1	2

b) 2-input NAND

NAND2				*N* transistor *A*						*N* transistor *B*						*P* transistor *B*						*P* transistor *A*					
A	B	I	Q	sd	gs	gd	sb	gb	db	sd	gs	gd	sb	gb	db	sd	gs	gd	sb	gb	db	sd	gs	gd	sb	gb	db
0	0	B	1	..	..	..	..	..	..	..	..	HX	..	..	HX	..	HX	HX	..	HX	..	..	HX	HX	..	HX	..
1	0	B	1	..	H.	H.	..	H.	..	HX	..	HX	..	..	HX	..	HX	HX	..	HX	..	..	..	..	..	..	..
0	1	B	1	HX	..	HX	..	..	HX	..	..	..	HX	HX	HX	..	..	..	..	..	..	..	HX	HX	..	HX	..
1	1	B	0	..	HX	HX	..	HX	..	..	HX	HX	..	HX	..	HX	..	HX	..	..	HX	HX	..	HX	..	..	HX
defect class				1	2	3	-	4	1	5	6	7	1	8	9	6	10	7	-	10	6	6	11	12	-	11	6

modeled behavior of this fault is high current and unknown logic value at output Q whenever input A is 1. This fault forms an equivalence class with the gate to bulk (gb) leakage on the N transistor, as well as the source to drain (sd) and drain to bulk (db) leakages on the P transistor. The use of unknown logic values is more general than attempting to determine actual values, and is both process and cell library independent. It also avoids restricting modeled detect behavior, which can reduce diagnostic effectiveness if actual behaviors are no longer accounted for. Experience has shown that the current readings and logical X propagation alone are often sufficient for accurate diagnosis.

Note that in some cases the logic value is known to be unchanged. An example is the gate to source (gs) short on N transistor A, in the 2 input NAND (Figure 2) when A is 1 and B is 0. Since N transistor B is off, the Q output of the gate is electrically isolated from the fault site.

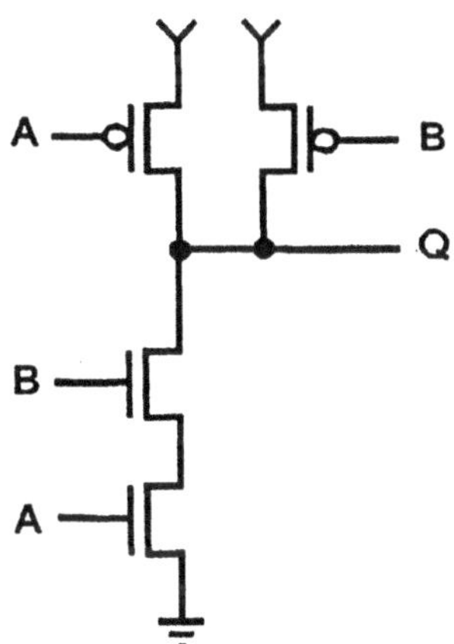

Fig. 2. 2 Input NAND gate.

In addition to the type of faults, their multiplicity and inter-relationships must also be considered when gates are combined into a circuit. A fault is *active* when it is drawing high I_{DDQ}, and its *activation* is defined to be the set of values on signal lines required to put it in its active state. Two faults are said to have *independent* activation if the presence of either cannot affect the activation of the other, otherwise the faults have *dependent* activation. Similarly, logical changes in the output cone emanating from the fault sites (*fault effects*) are independent if the output cones do not overlap. Using the concepts of activation and fault effects, four classes of multiple fault models, each a superset of the last, with increasing levels of diagnostic difficulty, are proposed:

I. Single fault.
II. Multiple faults: independent fault effects, independent activation.
III. Multiple faults: dependent fault effects, independent activation.
IV. Multiple faults: dependent fault effects, dependent activation.

Examples of circuit faults in each model are shown in Figure 3. Each box represents a circuit, inputs at the bottom and outputs at the top. An individual fault occurs at a single site, which has an associated input and output logic cone. In model II, for example, the input cones (I1 and I2) of two faults (1 and 2) may overlap, but their output cones (O1 and O2) remain independent. This permits results from single fault simulation of faults 1 and 2 to be merged for both detection and diagnosis. Model III permits the output cones to overlap, so the propagation of fault effects to primary outputs predicted by single fault simulation may be invalidated by other faults. Tests predicted to detect single faults with I_{DDQ} will still remain valid. Finally, in the model IV example the output cone of fault 1 forms part of the input cone to fault 2, and single fault simulation cannot guarantee detection of fault 2 nor aid diagnosis.

While each successive model will encompass more actual defects than its predecessor, it may be advantageous, depending on the application, to consider a simpler model. For example, typical scan circuits tend to be wide and flat. This means that many portions of the circuit are effectively independent during scan testing. This independence allows model II to be considerably more successful than it might be in tightly-coupled circuits like ALUs, where model III might be more appropriate. Accurate diagnosis using model IV virtually requires the use of effect-cause algorithms, as in [13], rather than fault simulation. Even these may break down when faced with indeterminate values and fault models such as those described above.

DIVA uses the transistor leakage fault model-described earlier for diagnosis under multiplicity models I through III. Despite the complexity of this diagnosis model, DIVA is able to generate tests for detecting modeled defects using only a modified single stuck-at fault model, as discussed in Section 3.

2.2. Previous Results

There are few previous results in fault diagnosis with current testing. The algorithm described in [8] is similar to DIVA without the voltage analysis. An exhaustive

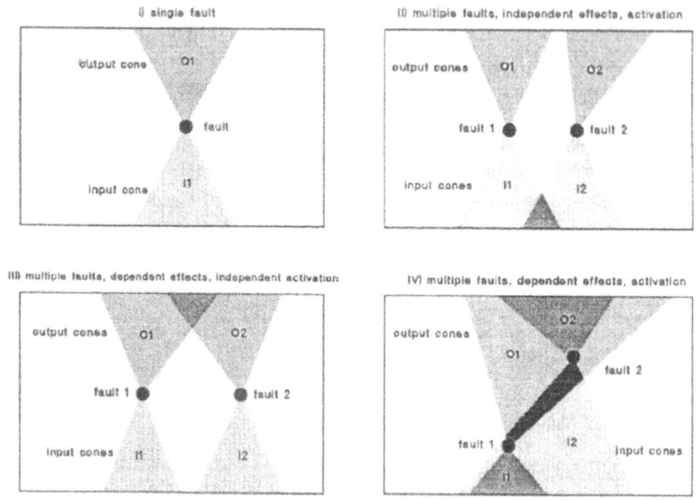

Fig. 3. Examples of circuit fault multiplicity models.

dictionary of leakage faults was generated, although there is conceptually no reason why a smaller test set could not be used. A diagnosis tool is alluded to in [3], but not described. DIVA itself was originally presented in [6] as an extension of the voltage-only methods described in [14] and [15]. A straightforward method for diagnosing single bridging faults is given in [9]. The method of [7] uses differential voltage contrast with images from electron beam probing; its defect model is thus visible differences between observed images and a *golden* set. As with other e-beam techniques, depassivated devices are preferred, and packaged parts must be decapped before using it.

A well-known diagnosis technique for large scan circuits and voltage testing alone is the intermediate signature method of Waicukauski et al [16]. This method uses signature analysis to isolate failures to blocks of 256 patterns. Several of these blocks must be stored to provide accurate diagnosis, and the method described in [16] will locate failures which behave as stuck-at faults at least part of the time under models I and II of the previous section. A major disadvantage of the method is that the failing blocks of 256 patterns may contain as few as 1 actual faulty vector. In contrast, the DIVA methodology stores only known faulty output vectors, and works on defects from models I through III, including some which are logically redundant.

Finally, there are the effect-cause methods of diagnosis, such as [13]. These are able to accurately diagnose faults under all multiplicity models of the previous section, but they typically require much larger test sets than those ordinarily generated. In addition, they assume a constancy in signal values which cannot be guaranteed in the presence of logic contention, or even between test vectors for a scan circuit whose flip-flop outputs are permitted to change while the scan chain is being loaded or unloaded (*destructive scan*).

3. Modifying Test Tools

In order to detect transistor leakages of the type described in Table 1, under multiplicity models I through III, a single stuck-at fault automatic test pattern generation (ATPG) tool may be altered as follows (results based on original work in [17]): All single stuck-at faults on gate inputs are to be targeted, and a fault will be listed as detected if its effect propagates to the gate's output, rather than a primary output. The resulting faults will be termed *pseudo stuck-at* faults, to distinguish them from conventional stuck-at faults, which must be propagated to a primary output.

Tests generated in this manner may be used to detect any leakage failures in simple gates (AND, NAND, OR, NOR, NOT) under models I through III. This claim may be verified by inspection for the gates in Table 1: For example, completely testing a 2-input NAND gate for input pseudo stuck-at faults requires the input patterns 11, 01, and 10, at least one of which detects every member of the 12 defect classes. For the result to apply to complex gates (including exclusive-ORs), they must be modeled by networks of simple (primitive) gates, and faults are listed as detected if they can be propagated to the network output, rather than the primitive gate output. It is also possible to model pads and flip-flops using simple gates, in order to mimic their functionality in scan test mode. Such models are highly dependent on test protocol and may or may not cover all defect classes within them. Activating some transistor shorts (those on clock lines, for example) in these elements may be impossible in scan test mode.

Existing test pattern generation algorithms for single stuck-at faults and voltage testing need only slight modification to generate tests required for fault diagnosis with DIVA. In order to perform fault diagnosis, some modifications are required to a conventional fault simulator. DIVA currently uses a parallel-pattern single fault propagation (PPSFP) simulator [18]. For verifying tests generated by the ATPG tool above, only fault-free circuit simulation is required, not fault simulation. For diagnosis, the simulator has been modified to propagate the X values to all primary outputs they reach, and to generate a fault dictionary.

4. The DIVA Diagnosis Algorithm

This section describes a demand-driven diagnosis procedure. If sufficient resources are available, a dictionary containing predicted responses of all modeled faults to all test patterns may be generated in advance. In that case, boolean operations, rather than fault simulation of the circuit, are performed on this dictionary to find faults which match a given behavior.

A parallel-pattern fault simulator (e.g., [18]) is also assumed. Diagnosis with a single-pattern simulator may tend to be somewhat slower. For the purposes of illustration, the word length of the diagnosing machine is taken to be 32.

Diagnosis proceeds in passes starting with an initial fault list and continuing until the list is reduced to a suitably small number of potential faults (ideally 1). The initial fault list may be either the entire set of modeled faults, or a reduced list. Parameters which may be used to reduce the list include, for example, the ability of the modeled fault to affect observed faulty outputs. Detailed techniques for using such parameters are available in [15].

The diagnosis algorithm in this section allows for multiple faults, but assumes that their activations are independent (models I through III of Section 2.1). This allows simulation to be performed on single faults and the results merged.

Effective diagnosis requires that sufficient failure information be available. Specialized hardware such as that discussed in Section 5 is perhaps the best way to obtain this information. Otherwise, test equipment of the Stop On First Error (SOFE) variety should be modified to record as many errors as possible.

For each diagnostic pass, a block of 32 patterns to simulate is required. This block may be formed by the observed failing patterns, a subset of them (if there are more than 32), or by the observed failing patterns padded with observed passing patterns (if there are less than 32). Including some passing patterns tends to speed diagnosis, since those faults predicted to fail on all patterns (often around 15%) may be eliminated quickly. This block of patterns is then simulated for each fault in the initial list. The simulation results are compared with the observed values and the initial fault list is divided into numbered groups as follows:

Let an m output circuit (define the I_{DDQ} reading as output 0) be simulated for n patterns, and define O_i^j as the observed value of output i for vector j, S_i^j as its simulated (predicted) fault-free value, and $F_i^j(d)$ as its simulated (predicted) faulty value in the presence of modeled fault d. The relation $\sim$ will be taken to mean *matches* and is defined such that $1 \sim 1$, $0 \sim 0$, $X \sim 1$, $X \sim 0$, and $X \sim X$ ($\nsim$ is the inverse of $\sim$). When a fault could belong to more than one group, it will be placed in the one with the lowest index number. Groups for original fault list:

1. Defects whose predicted output values match all observed output values. Thus, those d such that $F_i^j(d) \sim O_i^j$ for all i, j. This class comprises single modeled faults which completely explain the behavior of the circuit (model I of Figure 3).
2. Defects whose predicted output values match the observed output values on all outputs which can be reached from the fault site. Thus, those d such that $F_i^j(d) \sim O_i^j$ for all outputs i where a path exists from the site of d to i and all vectors j. This class encompasses those modeled faults in model II of Figure 3, whose fault effects are independent.
3. Defects whose predicted failures match observed output failures on all outputs which can be reached from the fault site. Thus, those d such that $F_i^j(d) \sim O_i^j$ whenever $F_i^j(d) \nsim S_i^j$ for all outputs i where a path exists from the site of d to i and all vectors j. These faults represent a partial explanation for observed failures, and behave as model II of Figure 3 for those patterns where they are predicted to fail (but not necessarily others).
4. Defects whose predicted failures match some observed output failures on outputs which can be reached from the fault site. Thus, those d such

there exist i and j such that $F_i^j(d) \sim O_i^j$ and $F_i^j(d) \not\sim S_i^j$.
These faults belong to model III (or possible IV) of Figure 3.
5. Defects whose predicted failures do not match any observed output failures. Thus, those d such that there do not exist i and j such that $F_i^j(d) \sim O_i^j$ and $F_i^j(d) \not\sim S_i^j$. Non-redundant faults in this group often tend to be in the immediate vicinity of faults in groups 1 and 2.

Defects in group 1 are the most likely to be the cause of the observed failure, while faults in group 5 cannot alone be the cause. Defects in the remaining groups may be the cause, with decreasing likelihood. Several faults together in group 2 may exactly explain the errors, as may some of those in group 3 whose effects partially overlap (although this would require more detailed simulation to verify). Defects in group 4 may be the cause, but only if they are acting intermittently, or are being affected in unknown ways by other faults.

After each pass through the algorithm, if additional resolution is desired, the most likely faults may be returned to the potential fault list, a new block of vectors taken (any remaining faulty vectors, or some which were observed to pass), and another pass may be run. Useful criteria for assigning likelihood are the number of *hits*, the number of i, j pairs where $F_i^j(d) \sim O_i^j$ and $F_i^j(d) \not\sim S_i^j$, and the number of *misses*, the number of i, j pairs where $F_i^j(d) \not\sim O_i^j$ and $F_i^j(d) \not\sim S_i^j$.

If, after a given pass, no faults remain as likely candidates, the pass may be repeated with some of the less likely candidates from the previous pass. When an unmodeled fault has occurred in a circuit, DIVA is likely to report some consistent modeled faults, typically in group 4 (although these are not guaranteed to be the cause).

The example which follows illustrates the groups and the algorithm.

4.1. Diagnosis Example

Consider the circuit of Figure 4. A set of four test vectors has been applied (these are labeled in pipeline fashion and should be read from right to left). High currents have been observed for vectors 1 and 4, and both outputs are in error for the fourth vector. Using the fault model shown in Table 1, the potential faults may be grouped as shown (a/b denotes fault class b within gate a):

1. 1/8, 1/12
2. 2/9
3. 1/1, 1/6, 1/11, 2/1, 2/5, 2/10, 2/11, 3/1, 3/11
4. 1/2, 1/3, 1/4, 1/7, 1/9, 2/2, 2/3, 2/4, 2/7, 2/8, 2/12, 3/2, 3/3, 3/4, 3/5, 3/7, 3/8, 3/9, 3/10, 3/12
5. 1/5, 1/10, 2/6, 3/6

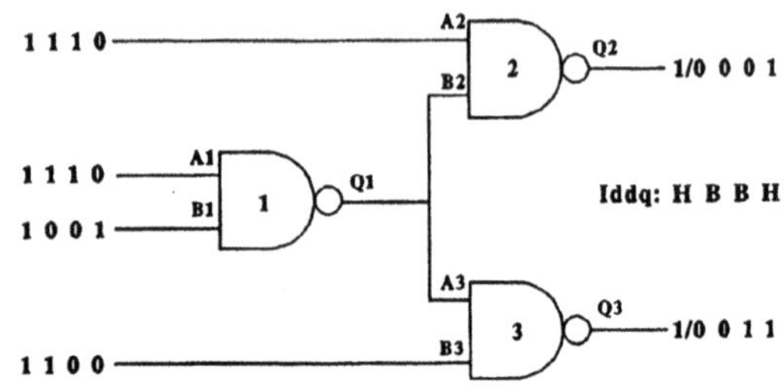

Fig. 4. Example circuit.

To generate the entry for 1/8, for example, first the input patterns on gate 1 are determined (01, 10, 10, and 11). Using Table 1b, the expected current (I) and gate output (Q) values are determined for class 8 (HX, $B1$, $B1$, and HX). The Q values are then propagated to the circuit outputs, resulting in 1, 0, 0, X for the output of gate 2 and 1, 1, 0, X for 3. Their fault-free values are 1, 0, 0, 1 and 1, 1, 0, 1 respectively, while the observed response is 1, 0, 0, 0 and 1, 1, 0, 0, with current readings of H, B, B, H. The predicted output values for 1/8 match the observed values for all vectors and outputs, so 1/8 is placed in group 1.

In this case, two modeled faults (1/8 and 1/12) precisely predict the observed behavior (all observed errors and high currents). To distinguish between them, the pattern 00 would need to be applied to gate 1. One fault on gate 2 explains the behavior of output $Q2$ and the high currents, but does not explain the error on output $Q3$ so belongs in group 2. 9 of the 36 modeled faults can account for some of the observed failures, but not all (group 3). Over half (20 of 36) of the modeled faults are consistent with some failures, but are inconsistent with some of the passing patterns (group 4). Finally, 4 of the 36 are completely inconsistent with all observations (group 5). In this case, it appears that 1/8 and 1/12 are the most reasonable fault hypotheses.

5. Diagnosis Hardware for DIVA

This section describes hardware which can be used to gather data for the diagnosis algorithm discussed in

Section 4. An ideal location for this circuitry would be as part of a (presumably self-testing) test control chip located on the board under test. This chip could be responsible for testing and gathering failure data from all other chips on the board. The functions could also be performed by a chip-specific test board, by a conventional tester, or even on-chip, although storing actual failing patterns would likely be impractical in the absence of substantial on-chip RAM.

The proposed circuitry is shown in Figure 5. Its basic operation is as follows: A test pattern generator produces input vectors for the circuit under test (CUT) (these may be generated pseudo-randomly, from ROM, or from an external file). The current drawn by the CUT is monitored, either on-chip (e.g., [2]) or externally. When a high current is detected, the corresponding output pattern (including scan chain values) along with the sequence number of the input pattern is stored. The test is terminated when the entire test set has been fed to the CUT, or when a predetermined number of failures (e.g., 32) have been observed. When the test is completed, the error storage area contains the sequence number of each vector where high current was observed, plus the actual output value for that vector. Because only known faulty patterns are stored, the storage requirements are reduced over those of the intermediate signature method of [16].

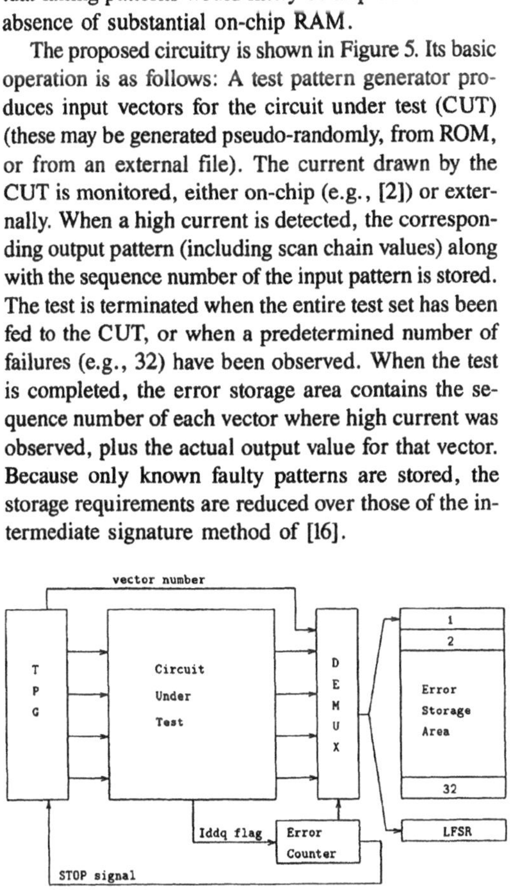

Fig. 5. Block diagram of diagnosis circuitry for DIVA.

One drawback of the circuitry of Figure 5 is that it will not record patterns where circuit outputs are incorrect, but no high current is drawn. For this reason, a linear feedback shift register (LFSR) is included to perform signature analysis on the CUT output. In the event that defects exist which alter the circuit outputs but do not draw increased current, this register will (neglecting aliasing) indicate a circuit failure. If diagnosis of these failures is desired, means other than those given in this article must be employed. It is assumed that few failures will fall into this category.

When the circuitry has recorded errors, the LFSR may be used to verify that these errors were the only ones which occurred. This is done by generating an input sequence for the LFSR which consists of the recorded error patterns inserted into the known fault-free sequence at the appropriate positions. The LFSR is then simulated over this sequence, and its final signature is compared with that observed in the hardware. If the two are equal, then no output logic errors occurred without a corresponding increase in I_{DDQ}. If not, incorrect logic values occurred without high I_{DDQ}, which may affect diagnostic accuracy and/or completeness.

6. Experimental Results

The utility of the DIVA algorithms is being evaluated on a standard cell ASIC. This chip is a full scan design containing approximately 36K transistors, and is fabricated in a 2-level metal 1 μm CMOS process. It has 101 inputs, 59 outputs, and a 436 element scan chain. Current monitoring is done off-chip. The I_{DDQ} test set for this chip consists of 140 scan vectors. Two experiments were performed on this chip. In the first, only the I_{DDQ} portion of the DIVA was used, while in the second both I_{DDQ} and voltage were used.

6.1. I_{DDQ}-only Failures

The chips selected for diagnosis in this experiment were those which fail the I_{DDQ} test, but pass all other production tests (functional, voltage scan, gross open/short, etc.). A sample of 151 chips was obtained for diagnosis. These parts represent about 1% of all those tested. Diagnosing these parts, which exhibit no logical failures, is in some sense a worst-case scenario for DIVA, since there are no logical output failures to guide diagnosis. On the other hand, since the chips have passed so many other tests, it is likely that each has only a single defect.

The distribution of failing currents of the 151 chip sample is given in Table 2. From the magnitude of these currents, it is clear that these chips contain defects (typical quiescent current on these chips is measured in nanoamperes).

Of the 151 chips in the sample, DIVA is able to diagnose 120 of them to one or more single leakage

Table 2. Distribution of failing currents.

Measured Current	Distribution	
	Absolute	Percentage
30–40 μA	11	7.3
40–60 μA	17	11.3
60–80 μA	9	6.0
80–100 μA	5	3.3
100–200 μA	9	6.0
200–300 μA	11	7.3
300–400 μA	11	7.3
400–500 μA	6	4.0
500–600 μA	9	6.0
600–800 μA	7	4.6
800–1000 μA	8	5.3
1–2 mA	16	10.6
2–3 mA	16	10.6
3–4 mA	8	5.3
4–5 mA	8	5.3

faults in group 1. The number of failing vectors varied from 1 to 140 (only one chip failed all 140 vectors), with two peaks, one at about 10 failing vectors and also at 70. The complete distribution is given in Figure 6. Total diagnosis time for all the chips was under 5 CPU minutes on an HP 9000 series 700 workstation, using a defect dictionary. Generating the dictionary takes about 5 CPU minutes on the same machine.

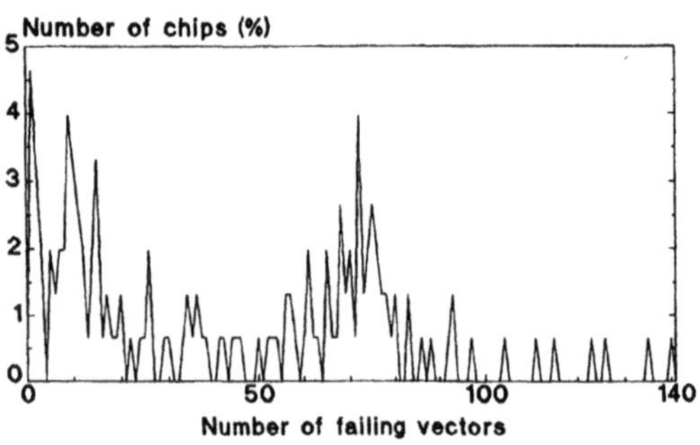

Fig. 6. Distribution of failing vectors (I_{DDQ} only).

Figure 7 shows the diagnostic resolution for the 151 sample chips. This figure shows that for 52 of the 120 chips for which transistor leakage diagnoses were obtained, the diagnosed defects were confined to a single standard cell (gate). Another 20 chips could have their behavior explained by a defect located in one of 2 gates, 13 chips could have defects in one of 3 gates, and so on. In addition to gates, figures are given for modules, which are one hierarchical level above standard cells. Modules are included to show that even when a fault could be in several gates, these are typically confined to a small portion of the circuit, often connected to one net.

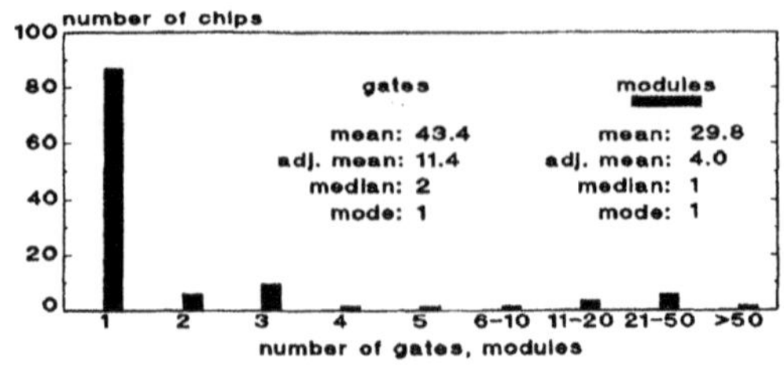

Fig. 7. Diagnostic resolution (I_{DDQ} only).

In addition to the absolute distribution, Figure 7 shows its mean, median, and mode. An *adjusted mean* is also given—this is the mean when the chip which failed all 140 vectors is excluded. The median number of modules is 1; that is, for the majority of the chips, their faulty behavior could be explained by faults which were confined to a single module, typically to 2 or fewer standard cells within that module. The means are heavily skewed by some large values. When all 140 vectors failed, any of 1649 gates in 895 modules could have been responsible. Of the other 10 chips where the diagnosis indicated that any of more than 50 gates could have been involved, 5 failed on one set of vectors, and 3 on another. In each case, the gates involved are connected to a global signal.

As of this writing, failure analysis has not been performed on these parts, so the diagnoses have not been verified.

6.2. I_{DDQ} and Voltage Failures

In the second experiment, defective chips with failing I_{DDQ} and voltage measurements were diagnosed using DIVA. Of an initial sample of 6 chips, 4 were diagnosed as having leakage faults (the other 2 had bridging faults).

3 of the 4 chips had a leakage which behaved as a single stuck-at fault during the voltage measurement. A gate to substrate leakage on the *N* (*P*) transistor connected to a given cell input is typically indistinguishable, using I_{DDQ}, from a stuck-at 0(1) fault on the net connected to the input. The I_{DDQ} measurements alone could thus identify the affected net, but were unable to determine the actual cell without the output voltage measurements.

No single defect could match the behavior of the final chip. As a result, multiple faults were investigated, using model II of Figure 3. The combined I_{DDQ} and voltage measurements showed that failures on 3 gates could together account for the failing vectors. Analysis of the circuit showed that the inputs associated with the predicted faults for each of the 3 gates were on fanout branches of a single line. The 4th branch was evidently fault free. Examination of the chip 2 using an electron microscope has identified a physical defect at the location predicted—a metal particle covering 3 of the 4 fanout branches.

Failure analysis on these chips have so far been limited to what can be seen in the metal layers. Defects within cells have so far not been visually confirmed. Instead, electron beam probing has verified that the predicted cells are defective without being able to identify the actual defects.

7. Conclusions

The DIVA method for performing diagnosis in combinational static CMOS circuits has been presented. The method resolves defects based on a model which includes single and several classes of multiple defects, which is believed to be more realistic than a conventional single stuck-at approach for scan circuits. Successful single leakage diagnoses have been obtained for about 80% of a sample of parts using I_{DDQ} alone. When both I_{DDQ} and voltage are used, the success rate is 100% for the parts tried, although fewer such parts have been investigated. DIVA's hardware and storage requirements are simple, and may be implemented as part of an on-board test control unit. Test generation for the method requires only minor modifications to existing single stuck-at ATPG software.

Acknowledgments

Assistance from Inshen Chiang, Rick Garvin, Midge James, Vic Johansen, Peter Maxwell, Garry O'Brien, Jeff Schoper, Bob Shreeve, and the JETTA reviewers is gratefully acknowledged by the author.

References

1. F.J. Ferguson, M. Taylor, and T. Larrabee, "Testing for parametric faults in static CMOS circuits," *Proc. ITC-90*, pp. 436–443, Washington, DC, September 1990.
2. W. Maly and P. Nigh, "Built-in current testing—feasibility study," *Proc. ICCAD-88*, pp. 340–343, Santa Clara, CA, November 1988.
3. W. Mao, R.K. Gulati, D.K. Goel, M.D. Ciletti, "QUIETEST: A quiescent current testing methodology for detecting leakage faults," *Proc. ICCAD-90*, pp. 280–283, Santa Clara, CA, November 1990.
4. P. Nigh and W. Maly, "Test generation for current testing," *IEEE Design and Test*, vol. 7, pp. 26–38, February 1990.
5. J. Soden and C. Hawkins, "Electrical properties and detection methods for CMOS IC defects," *Proc. European Test Conf.*, pp. 159–167, April 1989.
6. R.C. Aitken, "Fault location with current monitoring," *Proc. ITC-91*, pp. 623–632, Nashville, TN, October 1991.
7. R. Bottini, D. Calvi, S. Gaviraghi, A. Haardt, M. Vanzi, and D. Turrini, "Failure analysis of CMOS devices with anomalous IDD currents," *Proc. ISTFA 91*, pp. 381–388, Los Angeles, CA, November 1991.
8. D.J. Burns, "Locating high resistance shorts in CMOS circuits by analyzing supply current measurement vectors," *Proc. ISTFA-89*, Los Angeles, CA, November 1989.
9. S. Chakravarty and M. Liu, "Algorithms for current monitor based diagnosis of bridging and leakage faults," *Proc. DAC-92*, pp. 353–356, Anaheim, CA, June 1992.
10. J.M. Acken and S.D. Millman, "Accurate modeling and simulation of bridging faults," *Proc. CICC-91*, pp. 17.4.1–17.4.4, San Diego, CA, May 1991.
11. J.P. Shen, W. Maly, and F.J. Ferguson, "Inductive fault analysis of MOS integrated circuits, *IEEE Design and Test*, vol. 2, pp. 13–36, December 1985.
12. S.D. McEuen, "I_{DDq} benefits," *Proc. VLSI Test Symp.*, pp. 285–290, April 1991.
13. H. Cox and J. Rajski, "A method of fault analysis for test generation and fault diagnosis," *IEEE Trans. Comp. Aided Design*, vol. 7, pp. 813–833, July 1988.
14. R.C. Aitken and V.K. Agrawal, "A diagnosis method using pseudo-random vectors without intermediate signatures," *Proc. ICCAD-89*, pp. 574–577, Santa Clara, CA, November 1989.
15. R.C. Aitken, A Hierarchical Method of Fault Diagnosis with Built-In Self-Test Applications, Ph.D. Dissertation, Dept. of Electrical Engineering, McGill University, Montreal PQ, April 1990.
16. J.A. Waicukauski and E. Lindbloom, "Failure diagnosis of structured VLSI," *IEEE Design and Test*, vol. 6, pp. 49–60, August 1989.
17. Y.K. Malaiya and S.Y.H Su, "A new fault model and testing technique for CMOS devices," *Proc. ITC-82*, pp. 25–34, November 1982.
18. J.A. Waicukauski, E.B. Eichelberger, D.O. Forlenza, E. Lindbloom, and T. McCarthy, "A statistical calculation of fault detection probabilities by fast fault simulation," *Proc. ITC-85*, Philadelphia, PA, November 1985.

Robert C. Aitken received the B.Sc. (Honours) and M.Sc. degrees in computer science from the University of Victoria, Canada in 1985 and 1986 respectively, and the Ph.D. degree in Electrical Engineering from McGill University, Montreal, Canada in 1990.

Since 1990 he has been a member of the technical staff at Hewlett-Packard's Design Technology Center in Santa Clara, California. His research interests include fault diagnosis, fault modeling, design for testability, and built-in self-test. Dr. Aitken is a member of the IEEE.

JOURNAL OF ELECTRONIC TESTING: Theory and Applications, 3, 377-385 (1992)

Algorithms for I_{DDQ} Measurement Based Diagnosis of Bridging Faults*

SREEJIT CHAKRAVARTY**

Dept. of Computer Science, State University of New York, Buffalo, NY 14260

MINSHENG LIU***

Performance Technologies, Inc., 315 Science Parkway, Rochester, NY 14620

Received February 4, 1992. Revised August 10, 1992.

Editor: R.K. Gulati

Abstract. In the absence of information about the layout one is left with no alternative but to consider all *bridging faults*. An algorithm for diagnosis of a subset of such faults, viz. *single two line bridging faults* in static CMOS combinational circuits is presented. This algorithm uses results from I_{DDQ} *measurement* based testing.

Unlike known diagnosis algorithms, this algorithm *does not use fault dictionaries*, it uses *only logic simulation* and uses *no fault simulation*. It also uses *SOPS*, a novel representation of subsets of two-line *bridging faults* resulting in an efficient algorithm.

In spite of the large number of faults that we consider, our experimental results point to the computational feasibility of I_{DDQ} Measurement based diagnosis of *single two line bridging faults*. It also shows the effectiveness of reducing the set of possible faults using I_{DDQ} measurements.

Keywords: Bridging faults, diagnosis algorithm, I_{DDQ} testing.

1. Introduction

Recent studies have highlighted the importance of testing and diagnosing CMOS circuits for *bridging* and *leakage faults* [1]–[3]. As discussed in [4]–[6], I_{DDQ} measurements, using either built-in or external current sensors, can be used to detect such faults. This test methodology is known as I_{DDQ} *Testing*. The corresponding tests are called I_{DDQ} *Tests*. I_{DDQ} *Tests* and I_{DDQ} *Testing* are discussed in more detail in Section 2.

Algorithms for diagnosing *leakage faults* have been discussed in [7], [8]. Our algorithm for diagnosing *leakage faults* [8] is similar to the diagnosis algorithm in [7]. In [8] experimental evaluation of the algorithm on the ISCAS85 benchmark circuits [9] is also presented. In this article we focus our attention on *bridging faults*.

*A preliminary version of this work was presented at the 29th ACM/IEEE Design Automation Conference, 1992.

**Research Partially Supported by NSF Grant No. MIP-9102509.

***This work was performed when the author was with the Dept. of Computer Science, State University of New York at Buffalo.

1.1. Fault Model

Any given bridging fault is of the following form: there are sets of lines $S_1, \ldots, S_t$ where each S_i consists of two or more lines of the circuit; and for all S_i the lines in S_i are shorted. If there is only one such set we say that it is a *single bridging fault* else it is a *multiple bridging fault*. If for all j, $\|S_j\| \leq K$ then we say that the fault is a *K-line bridging fault*.

In this article, we assume that I_{DDQ} *Testing* has identified the circuit to be faulty and the faulty response of the circuit is due to only one *single bridging fault*. Moreover, only *two lines* are involved in the *bridging fault*. Henceforth we refer to such faults as *two-line single bridging faults* (**TSBFs**). Furthermore, we assume that any two lines in the circuit could be bridged and that, because of the fault, a very low resistance path exists between the two lines involved in the *bridging fault*.

We divide the set of *bridging faults* into two classes: *feedback* and *non-feedback* faults. If ∃ a path from X to Y then $<X, Y>$ is said to be a *feedback bridging fault* (**FBF**) between nodes X and Y. Else, $<X, Y>$ is a *non-feedback bridging fault* (**NFBF**). In Figure 1(a),

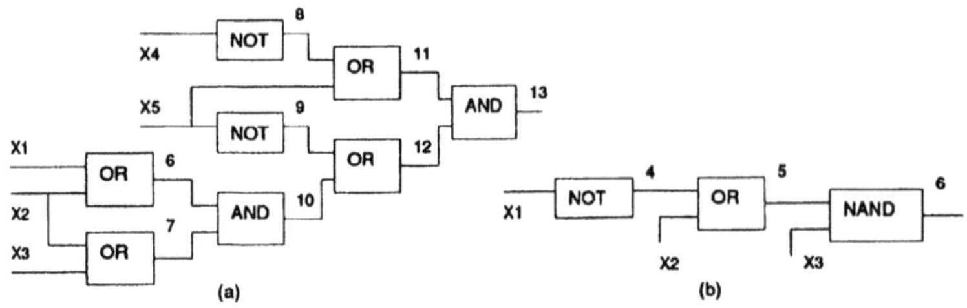

Fig. 1. Example circuits used for illustrating the algorithms.

<6, 12> is an FBF whereas <6, 7> is an NFBF. If the *bridging fault* is a TSBF then we use the abbreviations **FTSBF**, **NFTSBF** instead of **FBF**, **NFBF**.

1.2. Motivation for the Fault Model

Because of the large number of possible TSBFs in any circuit existing studies of *bridging faults* consider only a subset of all TSBFs. For instance, in [10] it is assumed that the layout of the circuit is available and all defects are a result of only one failure mode viz. *point defect*. The fault list is extracted by injecting such defects into the layout.

In our study we do not use this approach for the following reasons. Firstly, the layout of the circuit is not always available. Secondly, point defects are by no means the only failure mode. We therefore consider all TSBFs.

An important point about the work presented here is to show that it is computationally feasible to analyze the large number of faults in this fault model. However, to do that we need to get away from traditional techniques (used for stuck-at fault analysis) for designing the relevant computer-aided analysis tools. Here we illustrate the above point by presenting a novel technique for diagnosis of TSBFs using I_{DDQ} measurements.

1.3. Problem Definition

The *diagnosis problem* is defined as follows. Given a combinational circuit *C*, detected by an I_{DDQ} test set *T* to be faulty, identify the fault in *C*. We assume that the following information is available: (i) the gate level description of the fault free circuit; (ii) the definition of the input vectors in *T*; and (iii) for each input vector T_i in *T* if *C* gave a faulty response or not i.e., whether an abnormally high or a normal I_{DDQ} was measured. Furthermore, we assume that *C* has **only one** ***TSBF***. The objective is to determine what this fault is. Note that we do not take into account the logic response of the circuit.

1.4. Main Contribution

In this article algorithms for the *diagnosis problem* are described. We assume that the fault can be any TSBF.

Fault Dictionaries consist of a number of lists, one for each input vector in the test set. The list for an input vector T_i consists of the set of faults detected by it. Known diagnosis algorithms for *stuck-at* and *leakage faults* use *fault dictionaries*. The problem of using such an approach for bridging faults is the following. If *m* is the size of the circuit and *N* the size of the test set then such algorithms have space complexity $O(m^2N)$. Consequently, we do not use fault dictionaries. Instead, a **set of ordered-pair of sets (SOPS)** is used to represent the set of TSBFs. It requires $O(m)$ storage to represent the $O(m^2)$ bridging faults thereby resulting in a space efficient algorithm.

All diagnosis algorithms essentially perform the following task. They start with an initial set of *possible faults*. Then faults are dropped from this set by processing each input vector in the test set and the response of the circuit to the input vector. The process stops once the test set has been exhausted. Conventional diagnosis algorithms use the fault dictionary to drop faults. In our algorithm, linear time operations for dropping faults from fault lists represented by SOPS are used. This results in efficient diagnosis algorithms for bridging faults in combinational circuits.

Another important point about our diagnosis algorithm is that it uses only logic simulation and no fault simulation. We present two algorithms for combinational circuits. One of these is *optimistic* while the other is *pessimistic*. These terms will become clear in Section 3.

Experimental results, using the ISCAS85 [9] benchmark circuits, are presented in Section 4. These results do point to the computational feasibility of diagnosing all *TSBFs* in combinational circuits using I_{DDQ} *measurements*. Moreover, they also point to the effectiveness in reducing the set of *possible TSBFs* using I_{DDQ} *measurements*.

1.5. Related Work

Diagnosis algorithms for a variety of fault models have been discussed. The diagnosis algorithms in [11]–[16], assume that the faults are *stuck-at faults* and input/output experiments are used to detect the presence of faults. The work presented in [7] is for *leakage faults*, in combinational or LSSD based circuits, when I_{DDQ} *measurements* are used.

One of the earliest algorithms for I_{DDQ} measurement based diagnosis of bridging faults is discussed in [17]. This algorithm is similar to DIAG-BFC-OPT discussed in Section 3. If m is the number of nodes in the circuit and N the number of test vectors then this algorithm uses $O(mN)$ space and its worst case run time is $O(m^2N)$. In contrast, DIAG-BFC-OPT uses $O(m)$ space and its worst case run time is $O(mN)$. Moreover, as discussed in Section 3, unlike DIAG-BFC-PESS (which is the algorithm that we are proposing) both the algorithm in [17] and DIAG-BFC-OPT will incorrectly diagnose some classes of bridging faults.

The rest of the article is organized as follows. In Section 2 we discuss in greater details I_{DDQ} *Tests* for *bridging faults* and I_{DDQ} *Testing*. In Section 3 we present the algorithms. Section 4 contains the experimental results. Complexity bounds of these algorithms are discussed in Section 5.

2. I_{DDQ} Testing and I_{DDQ} Tests

Every static CMOS gate consists of two networks—a *pullup* network of *pFETs* and a *pulldown* network of *nFETs*. If X, the output of a gate, is set to 1(0) then there is a conducting path from V_{DD} (X) to X (V_{SS}). On application of any input vector, in the steady state and in the fault free condition, no current flows through the two FET networks. Therefore the quiescent current, denoted by I_{DDQ}, in the fault free circuit is small enough to be negligible. If there is a *bridging fault*, it can be excited, as discussed below, so that I_{DDQ} is large. Current sensors, either onchip or offchip, can be used to detect the large I_{DDQ} and can therefore be used to detect the presence of *bridging faults*. Such a test methodology is known as I_{DDQ} *Testing*.

Proposition 2.1, Proposition 2.2, stated below, are two sets of criteria used to determine if an input vector T is an I_{DDQ} *Test* for a given TSBF. We prefer to use Proposition 2.1 rather than Proposition 2.2 [5]. As can be seen from the discussion below, Proposition 2.2 is a much weaker criteria to be satisfied specially for FTSBFs. Also, we will see why we are insisting on a much stronger criteria for FTSBFs. For any line Z, let $T(Z)$ be the logic value at Z on application of T to the fault free circuit.

Proposition 2.1. T is an I_{DDQ} test for the NFTSBF $<X, Y>$ iff $T(X) \neq T(Y)$. Let $<X, Y>$ be an FTSBF s.t. there is a path from X to Y and $T'(Y)$ be the value at Y on application of T obtained by setting X to U (don't care) and using 3-valued simulation to determine the value at Y. Then, T is a test for the FTSBF $<X, Y>$ iff $T'(Y) \neq u$ and $T(X) \neq T'(Y)$.

Proposition 2.2. T is an I_{DDQ} test for the *TSBF* $<X, Y>$ iff $T(X) \neq T(Y)$.

First, by means of an example, we show why the condition $T(X) \neq T(Y)$ is enough to guarantee that T is a test for the NFTSBF $<X, Y>$. For the circuit of Figure 1(a), let $T1 = <X1 = 0, X2 = 0, X3 = 1, X4 = 0, X5 = 1>$. First consider the fault free circuit. We know that $T1(6) = 0$ and $T1(7) = 1$. Therefore, there is a conducting path $q1$ from V_{DD} to 7 and another conducting path $q2$ from 6 to V_{SS}. Next, consider the faulty circuit with the NFTSBF $<6, 7>$ in it. In the faulty circuit, we assume that the values at 6, 7 are unknown because they depend on the conductances of the pFETs and nFETs. These conductance values are not known because we have no knowledge of the FET sizes. However, since the fault is an NFTSBF there is no path from 6 (7) to the portion of the circuit that drives gate 7 (6), i.e., $X2$, $X3$ ($X1$, $X2$). Therefore, the values at $X1$, $X2$, $X3$ in the faulty circuit, on application of $T1$, are the same as their fault free logic values. Therefore, on application of $T1$, the conducting paths $q1$, $q2$ also exist in the faulty circuit. Hence, in the faulty circuit, on application of $T1$, there is a low resistance path from V_{DD} to V_{SS} (through $q1$, the fault $<6, 7>$ and q_2). Therefore, on application of $T1$ to the faulty circuit there will be a large I_{DDQ}. Thus, $T1$ is a test for the fault $<6, 7>$.

Next, we illustrate with the help of an example why the condition $T(X) \neq T(Y)$ (as stated by Proposition 2.2) is not sufficient to declare T a test for an FTSBF $<X, Y>$. This also explains why we prefer Proposition

2.1 to Proposition 2.2. Refer to Figure 1(b). Let $T2 = <X1 = 1, X2 = 0, X3 = 1>$. First consider the fault free circuit. We have $T2(4) = 0$, $T2(6) = 1$ and there is a conducting path $q1$ from 4 to V_{SS} and a conducting path $q2$ from V_{DD} to 6. Next consider the faulty circuit with the FTSBF $<4, 6>$ in it. As before, we have to assume that the logic values at 4, 6 on application of $T2$ are ambiguous. Since the values at $X2$, $X3$ are not dependent on the value at 4 therefore in the faulty circuit lines $X2$, $X3$ will have the logic value 0, 1 respectively. However, in the faulty circuit the value at 5 will be ambiguous. Consequently, there may or may not be a conducting path $q2$ from V_{DD} to 6. Hence $T2$ may not activate a low resistance path from V_{DD} to V_{SS}. Therefore, although $T2(4) \neq T2(6)$ it is not a test for the FTSBF $<4, 6>$. Hence, Proposition 2.1 would reject $T2$ as a test for $<4, 6>$ but Proposition 2.2 will accept it. Herein lies the difference between Proposition 2.1 and Proposition 2.2. Next consider $T3 = <X1 = 1, X2 = 0, X3 = 0>$. The reader can verify that $T3$ will be a test for the FTSBF $<4, 6>$ according to Proposition 2.1.

3. Diagnosis Algorithms

We first present an example and then formally state our first algorithm. Later on we will point out some shortcomings of this algorithm and suggest modifications to get a second algorithm. Experimental evaluation of these algorithms is presented in Section 4.

3.1. Example

The basic idea used by our algorithms is illustrated with the help of the circuit of Figure 1(a) and the Test Set and Test Response of Table 1. An entry 1(0) in the last column indicates if there was a large (small) I_{DDQ} on application of the corresponding test. Table 1 also shows, for each input vector T_i, the values at the lines on application of T_i.

We use **SOPS**, a compact representation of TSBFs. SOPS is a set of ordered pairs of the form $\{<A_1, B_1>, <A_2, B_2>, \ldots, <A_k, B_k>\}$ where: A_i is a set of lines in the circuit and B_i can be either a set of lines in the circuit or the special symbol Ψ. If B_i is a set of lines then the ordered pair $<A_i, B_i>$ denotes the set of TSBFs $<X, Y>$ where $X \in A_i$, $Y \in B_i$. If B_i is the special symbol Ψ then $<A_i, \Psi>$ denotes the TSBFs $<X, Y>$ where $X, Y \in A_i$. For example, the SOPS $L = \{<\{2, 3\}, \{7, 8\}>, <\{9, 17, 18\}, \Psi>\}$ represents the following set of TSBFs: $\{<2, 7>, <2, 8>, <3, 7>, <3, 8>, <9, 17>, <9, 18>, <17, 18>\}$.

The algorithm initially assumes that all TSBFs are possible. For the circuit of Figure 1(a), this set of TSBFs is represented by the *SOPS* $L_0 = \{<\{1, 2, \ldots, 13\}, \Psi>\}$. Next the input vectors are processed and faults are dropped from the current set of possible faults to get a new set of possible faults. Let L_i be the *SOPS* denoting the possible set after processing the input vector T_i. In this example we will use the weaker condition given by Proposition 2.2. Later in Section 3.3 we will modify the algorithm to conform with Proposition 2.1.

Let A be the set of all lines in the circuit and $B(C)$ the subset of A that is set to 0(1) on application of $T1$. From Table 1, we have: $B = \{1, 2, 3, 4, 5, 6, 7, 10\}$; and $C = \{8, 9, 11, 12, 13\}$. Since $T1$ did not give a faulty response the fault must be between the lines in B or between the lines in C but not between a line in C and a line in B. These set of TSBFs is represented by the SOPS

$$L_1 = \{<B, \Psi>, <C, \Psi>\}$$
$$= \{<\{1, 2, 3, 4, 5, 6, 7, 10\}, \Psi>, <\{8, 9, 11, 12, 13\}, \Psi>\}$$

Next, we process $T2$. Let $D(E)$ be the subset of B set to 0(1) by $T2$ and $F(G)$ the subset of C set to 0(1) by $T2$. From Table 1, we have $D = \Phi$; $E = \{1, 2, 3, 4, 5, 6, 7, 10\}$; $F = \{8, 9\}$; and $G = \{11, 12, 13\}$. From Table 1, we know tht there was a faulty response

Table 1. Logic simulation for our example.

	X1	X2	X3	X4	X5	6	7	8	9	10	11	12	13	R
T1	0	0	0	0	0	0	0	1	1	0	1	1	1	0
T2	1	1	1	1	1	1	1	0	0	1	1	1	1	1
T3	0	1	0	1	0	1	1	0	1	1	0	1	0	1
T4	1	0	0	1	1	1	0	0	0	0	1	0	0	0
T5	1	1	1	0	0	1	1	1	1	1	1	1	1	0

on application of $T2$. Therefore, the fault must be between a line in D and a line in E or a line in F and a line in G. Therefore, after eliminating ordered-pairs representing empty sets we have the following SOPS. This new SOPS represents the reduced set of possible TSBFs.

$$L_2 = \{<F, G>\} = \{<\{8, 9\}, \{11, 12, 13\}>\}$$

Next $T3$ is processed. Let $H(J)$ be the subset of F set to 0(1) by $T3$ and $K(M)$ the subset of G set to 0(1) by $T3$. From Table 1, we deduce that $H = \{8\}$; $J = \{9\}$; $K = \{11, 13\}$; and $M = \{12\}$. From Table 1 we know that there was a faulty response on application of $T3$. Therefore, the fault must be between a line in H and a line in M or a line in J and a line in K. From this we get the *SOPS* L_3.

$$L_3 = \{<H, M>, <K, J>\}$$
$$= \{<\{8\}, \{12\},>, <\{11, 13\}, \{9\}>\}$$

Continuing this way, after processing $T5$ we have:

$$L_5 = \{<\{8\}, \{12\}>, <\{13\}, \{9\}>\}$$

From this we deduce the possible set of bridging faults to be <8, 12>, <13, 9>. Thus from a possible set of 78 bridging faults we have narrowed it down to 2 bridging faults.

3.2. *Algorithm DIAG-BFC-OPT*

Notations.

N is the number of input vectors in the test set.
$T = \{T_1, \ldots, T_N\}$ is the test set.
$R = \{r_1, \ldots, r_N\}$ is the response vector where $\forall i$, $r_i \in \{0, 1\}$ and r_i = 0(1) if there wasn't (there was) a large I_{DDQ} on application of T_i.
C is the circuit being diagnosed.
X is the set of all lines in C.
m is the number of gates in C.
$T_i(y)$ is the value at line y on application of T_i.
n is the number of primary inputs of C.
$\{X_1, \ldots, X_n\}$ is the set of primary inputs of C.
$A^j(T_i)$ is the subset of A s.t. $\forall y \in A^j(T_i)$, $T_i(y) = j$, $j \in \{0, 1\}$.
$<A, \Psi>$ is the set of pairs of lines (y_1, y_2) s.t. $y_1, y_2 \in A$.
$<A, B>$, $B \neq \Psi$, is the set of pairs of lines (y_1, y_2) s.t. $y_1 \in A$ and $y_2 \in B$.
L_i is the set of pairs of lines, denoting the set of possible faults deduced by the algorithm, after it has processed vectors $T_1, \ldots, T_i$.

Step 1. $L_0 = \{<X, \Psi>\}$.

Step 2. For each T_i, $1 \leq i \leq N$ do Step 3.

Step 3. let $L_{i-1} = \{<A_1, B_1>, \ldots, <A_q, B_q>\}$. For $1 \leq j \leq q$ process $<A_j, B_j>$ to form new ordered pairs as follows. These new ordered pairs are added to L_i and $<A_j, B_j>$ is deleted from L_{i-1}.

Compute $A_j^0(T_i)$, $A_j^1(T_i)$. If $B_j \neq \Psi$, compute $B_j^0(T_i)$, $B_j^1(T_i)$. We have two cases.

Case 3.1. $B_j = \Psi$. If $r_i = 0$ then the new ordered pairs are $<A_j^0(T_i), \Psi>$ and $<A_j^1(T_i), \Psi>$. If $r_i = 1$ then the new ordered pair is $<A_j^0(T_i), A_j^1(T_i)>$.

Case 3.2. $B_j \neq \Psi$. If $r_i = 0$ then the new ordered pair(s) are $<A_j^0(T_i), B_j^0(T_i)>$ and $<A_j^1(T_i), B_j^1(T_i)>$. If $r_i = 1$ then the new ordered pair(s) are $<A_j^0(T_i), B_j^1(T_i)>$ and $<A_j^1(T_i), B_j^0(T_i)>$.

Note that if either A or B is empty then $<A, B>$ denotes an empty set of faults and is therefore not added to the list L_i. Similarly, if A is empty then $<A, \Psi>$ denotes an empty set and is not added to L_i.

Step 4. Let $L_N = \{<C_1, D_1>, \ldots, <C_p, D_p>\}$. The union of the set of bridging faults represented by these ordered pairs is the possible set of bridging faults.

3.3. *Algorithm DIAG-BFC-PESS*

Let us have a closer look at DIAG-BFC-OPT. A vector T_i is **good (faulty) vector** iff the corresponding *response bit* r_i is a 0(1). In Table 1, $T1$ ($T2$) is a *good (faulty) vector.*

Fact 1. While processing *good vectors* DIAG-BFC-OPT drops sets of faults of the form $<A_j^0, A_j^1>$ (Case 3.1) or $<A_j^0, B_j^1>$, $<A_j^1, B_j^0>$ (Case 3.2).

Fact 2. While processing *faulty vectors* DIAG-BFC-OPT drops sets of faults of the form $<A_j^0, \Psi>$, $<A_j^1, \Psi>$ (Case 3.1) or $<A_j^0, B_j^0>$, $<A_j^1, B_j^1>$ (Case 3.2).

Since the only process by which DIAG-BFC-OPT drops faults from the current set of possible faults are by the process described in the above two facts and both these are consistent with Proposition 2.2 we have the following.

Observation 3.1. DIAG-BFC-OPT is consistent with Proposition 2.2.

Next we observe that faults dropped by the process described by Fact 1 is not consistent with Proposition 2.1. This is because, in the process of dropping faults described by Fact 1 it is assumed that T detects a TSBF $<X, Y>$ iff $T(X) \neq T(Y)$. However, as explained in Section 2, many FTSBFs $<X, Y>$ may not give a faulty response on application of T_i even if $T_i(X) \neq T_i(Y)$. Therefore, the process of dropping faults as described by Fact 1 is consistent with Proposition 2.1 **only for NFTSBFs but not for FTSBFs**. Also, the process of dropping faults as described by Fact 2 assumes that if $T(X) = T(Y)$ then T does not detect the TSBF $<X, Y>$. Therefore, the process of dropping faults as described by Fact 2 is consistent with Proposition 2.1. We thus have the following observations.

Observation 3.2. The processing of *faulty vectors* by DIAG-BFC-OPT is consistent with Proposition 2.1.

Observation 3.3. The processing of *good vectors* by DIAG-BFC-OPT is consistent with Proposition 2.1 for NFTSBFs but not for FTSBFs.

The above two observations are combined to give us **Algorithm DIAG-BFC-PESS** which is consistent with Proposition 2.1.

Algorithm DIAG-BFC-Pess

/* The algorithm consists of two phases. *Phase I* consists of Steps 1, 2 and 3. *Phase II* consists of Step 4.

This algorithm first computes two sets of TSBFs $PS1$, $PS2$. $PS1$ contains the necessary NFTSBFs whereas $PS2$ contains the necessary FTSBFs. These two sets are then merged to get the desired set of possible faults. */

Step 1. $PS1$ is computed as follows. It is initialized to the *SOPS* $\{<X, \Psi>\}$ where X is the set of all lines in the circuit. Faults are then dropped from this initial set by processing all the vectors, *good* as well as *faulty*, as described in DIAG-BFC-OPT.

Step 2. $PS2$ is computed as follows. Like $PS1$, it is initialized to the *SOPS* $\{<X, \Psi>\}$ where X is the set of all lines in the circuit. Faults are then dropped from this initial set by processing only the *faulty vectors*, and none of the *good vectors*, as described in DIAG-BFC-OPT.

Step 3. $PS1$, $PS2$ are merged as follows. Let $FTSBF(PS2)$ be the set of FTSBFs in $PS2$. Then $PS = PS1 \cup FTSBF(PS2)$.

Step 4. If $\|PS\|$ is not small enough then more test vectors are generated. Let this additional test set be T'. Reduce $PS1$ by processing the *good* and *faulty* vectors in T' to get $PS1'$. Also, reduce $PS2$ by processing only the *faulty vectors* in T' to get $PS2'$. Then, our final set of possible faults PS' is given by $PS' = PS1' \cup FTSBF(PS2')$.

END(DIAG-BFC-PESS)

We note that Steps 1, 2 can be executed concurrently thereby saving on the number of logic simulations required and hence the total run time of the algorithm. We next illustrate DIAG-BFC-PESS with the example of Section 3.1. As computed in Section 3.1, $PS1 = \{<8, 12>, <9, 13>\}$. This is what DIAG-BFC-PESS will also compute. Note that $<9, 13>$ is a FTSBF and is included in $\|PS1\|$. This is because, as can be seen from Table 1, $T_i(9) = T_i(13)$ if T_i is a *faulty vector* and $T_i(9) \neq T_i(13)$ if T_i is a *good vector*.

To compute $PS2$ we start with $PS2 = \{<\{1, 2, \ldots, 13\}, \Psi>\}$. Since $T2$, $T3$ are the only two *faulty vectors* these are the only two vectors processed by DIAG-BFC-PESS. After processing $T2$ we have:

$$PS2 = \{<\{8, 9\}, \{1, 2, 3, 4, 5, 6, 7, 10, 11, 12, 13\}>\}.$$

After processing $T3$ we have:

$$PS2 = \{<\{9\}, \{1, 3, 5, 11, 13\}>, <\{8\} \{2, 4, 6, 7, 10, 12\}>\}$$

$FTSBF(PS2) = \{<5, 9>, <9, 13>, <4, 8>\}$. Therefore, PS computed by DIAG-BFC-PESS is given by: $\{<8, 12>, <9, 13>, <5, 9>, <4, 8>\}$. In the above set we note that $<5, 9>$ and $<4, 8>$ are FTSBFs that can never be dropped by our algorithm because, according to the conditions of Proposition 2.1, they are irredundant. In that sense DIAG-BFC-PESS is a pessimistic algorithm.

Having reduced the set of possible faults to a small number, more faults can be eliminated by doing fault simulation.

4. Experimental Results

In this section we present experimental results obtained by running the DIAG-BFC-PESS on ISCAS85 [9] combinational benchmark circuits. For each circuit we generated 25 instances of the diagnosis problem as follows. Random number generators were used to generate the numbers $n1$, $n2$ to get the *TSBF* $<n1, n2>$. The test vectors were also generated using a random number generator. As each T_i was generated, fault simulation was used to generate the *response bit* r_i using Proposition 2.1.

We used two *figures of merit* to evaluate the algorithm. The first was the *time required* and the

Table 2. Experimental results.

Name	Limit	Total	Length	Time	Size
c880	25	97903	80	0.624	172
c7552	100	6913621	90	6.845	1737
c6288	100	2995128	80	3.284	204
c60	15	703	30	0.044	17
c5315	100	3086370	690	24.378	5351
c499	25	29403	425	1.153	570
c432	25	19110	130	0.565	87
c3540	75	1476621	450	7.411	1389
c2670	50	1127251	80	2.340	735
c1908	75	416328	865	7.314	952
c1355	50	171991	520	2.758	359

second was the *final size* of the set of *possible faults*. These results are tabulated in Table 2 and are explained below. These values are average values over the 25 instances that were generated for each circuit.

For each instance of a circuit a test set (Test 1) was generated. Test1 is used in *Phase I* and equals $\{T_1, \ldots, T_N\}$ where N is the number of vectors needed to reduce $\|PS1\|$ below a *critical value*. This *critical value* is shown as *Limit* in Table 2. In addition to using "$\|PS1\| < Limit$" as a terminating condition we also terminated *Phase I* if the number of consecutive vectors for which $\|PS1\|$ did not change reached 50.

If after *Phase I* $\|PS\|$ was not small enough we generated an additional test set Test2 = $\{T_{N+1}, \ldots, T_{N+M}\}$ where $M = 4 * N$. This test set was also generated using random numbers. The average value of $M + N$, i.e., the total test length, is shown as *Length* in Table 2.

The final value of $\|PS\|$ computed by DIAG-BFC-PESS is given by *Size* and the total time required, in CPU secs on a SUN Sparkstation 2, is given by *Time*. We compare *Size* with *Total*, which is the total number of TSBFs in the circuit. Since *Size* << *Total* we conclude that DIAG-BFC-PESS is very effective in reducing the number of possible faults. The CPU time required points to the computational feasibility of diagnosing the circuit for all *TSBFs*. The small number of faults remaining can be further reduced using fault simulation.

One should bear in mind that the values of *Length*, *Time*, *Size* shown in Table 2 are average values. The *standard deviation* was high in all cases although the *median* was much lower than the average. This is because the *Length*, *Time*, *Size* are very sensitive to the fault and the test set in question.

5. Complexity Bounds

The algorithms can be implemented very efficiently using the data structure shown in Figure 2, except for the computation of FTSBF(PS2) and the set union operation.

Information about all the nodes is stored in an array *NodeList*. There is one entry in *NodeList*, named *NodeListEntry*, for each node in the circuit. After logic simulation of T_i, *NodeListEntry* for node X contains $T_i(X)$.

L_{i-1} is an *SOPS* represented as a linked list of ordered-pairs $<A_j, B_j>$. Each A_j, B_j is itself a linked list consisting of a list of pointers to the nodes that are in that set.

After logic simulation A_j^0, A_j^1, B_j^0, B_j^1 can be computed by traversing the two lists A_j, B_j and inspecting

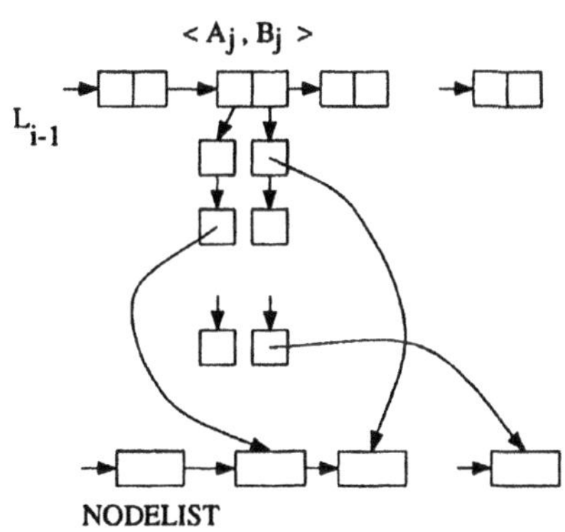

Fig. 2. Schematic of the data structure.

the values assigned by logic simulation. This takes time $O(\|A_j\| + \|B_j\|)$. From this the appropriate ordered pairs to replace $<A_j, B_j>$ can be computed in constant time. Therefore, if m is the number of nodes in the circuit, computing L_i from L_{i-1} takes time $O(m)$. This is because $\Sigma_{j=1}^{n} \|A_j\| + \|B_j\| \leq m$, where n is the number of ordered-pairs in L_{i-1} and every node can belong to only one set. Since logic simulation takes time $O(m)$ DIAG-BFC-OPT takes time $O(mN)$ where N is the number of vectors. Its space complexity is $O(m)$. FTSBF(PS2) is an expensive operation. But, $\|PS2\|$ is small enough that for all practical purposes the time complexity of DIAG-BFC-PESS is also $O(mN)$. Its worst case run time is $O(m^2N)$.

6. Conclusion

The main theme of this article is that by using SOPS, the quadratic number of bridging faults can be represented using linear space. Moreover, this efficient representation can be used to design an entirely new class of efficient diagnosis algorithms for diagnosing bridging faults. These algorithms for combinational circuits were experimentally shown to be very effective. The algorithms for combinational circuits can be modified to derive algorithms for sequential circuits with full scan.

Another issue of importance here is the fault model being used. Because of the large number of TSBFs researchers have been avoiding the use of such a fault model. Our work here suggests that the large number of faults being considered should not deter attempts to analyze such faults. By getting away from traditional techniques (those used for stuck-at faults) for diagnosis we were able to develop techniques that make I_{DDQ} measurement based diagnosis of all TSBFs in combinational circuits computationally feasible. One reason for this was the use of SOPS to represent subset of TSBFs. Such a representation can also be used for simulating and generating I_{DDQ} tests for all *TSBFs* [18]. Using this technique, we show experimentally, that like diagnosis fault simulating all TSBFs and generating tests for all TSBFs is also computationally feasible. This work along with the work in [18] shows that it is computationally feasible to do I_{DDQ} based analysis of all TSBFs in combinational circuits. It would be interesting to see if it is computationally feasible to analyze all TSBFs in sequential circuits.

Acknowledgments

We would like to thank the JETTA reviewers for suggestions they made to improve the presentation of the material.

References

1. F.J. Ferguson and J.P. Shen, "Extraction and simulatio of realistic faults using inductive fault analysis," in *Proc. 1988 Int. Test Conf.*, pp. 475–484, September 1988.
2. F.J. Ferguson and J.P. Shen, "A CMOS fault extractor for inductive fault analysis," *IEEE Trans. on Comput.-Aided Design*, vol. 7, pp. 1181–1194, November 1988.
3. J. Soden and C. Hawkins, "Test considerations for gate oxide shorts in CMOS ICs," in *IEEE Design & Test of Comput.*, pp. 56–64, August 1986.
4. D. Feltman, P.J. Nigh, L.R. Carley and W. Maly, "Current sensing for built-in testing of CMOS circuits," *1988 Int. Conf. on Computer-Aided Design*, pp. 454–457, Rye Brook, NY, October 1988.
5. P. Nigh and W. Maly, "Test generation for current testing," *IEEE Design and Test of Computers*, pp. 26–38, February 1990.
6. W. Mao, R.K. Gulati, D.K. Goel, M.D. Ciletti, "QUIETEST: A quiescent current testing methodology for detecting leakage faults," in *Proc. 1990 IEEE Int. Conf. on Computer-Aided Design*, pp. 289–283, November 1990.
7. R.C. Aitken, "Fault location with current monitoring, *IEEE Int. Test Conf.*, pp. 623–632, October 1991.
8. S. Chakravarty and M. Lu Liu, "Algorithms for current monitor based diagnosis of bridging and leakage faults," Tech. Rep. No. 92-05, Department of Computer Science, State University of New York, Buffalo, NY 14260.
9. F. Brglez and H. Fujiwara, "A Neutral netlist of 10 combinational benchmark circuits and a target translator in fortran," special session on ATPG and Fault simulation, *IEEE Int. Symposium on Circuits and Systems*, 1985.
10. F.J. Ferguson and T. Larabee, "Test pattern generation for realistic bridging faults in CMOS ICs," *IEEE Int. Test Conf.*, pp. 492–499, October 1991.
11. M. Abramovici, "A hierarchical, path-oriented approach to fault diagnosis in modular combinational circuits," *IEEE Trans. on Comput.*, vol. C-31, pp. 672–677, July 1982.
12. M. Abramovici and M.A. Breuer, "Multiple fault diagnosis of MOS Combinational Networks," *IEEE Trans. on Comput.*, vol. C-29, pp. 451–460, June 1980.
13. H. Cox and J. Rajski, "A method for fault analysis for test generation and fault diagnosis," *IEEE Trans. on Comput.-Aided Design*, vol. 7, pp. 813–833, July 1988.
14. Y.M. El-Ziq and S.Y.H. Su, "Fault diagnosis of MOS combinational networks," *IEEE Trans. on Comput.*, vol. C-31, February 1982, pp. 129–139.
15. J. Richman, K. Bowden, "The modern fault dictionary," in *Proc. Int. Test Conf.*, pp. 696–702, 1985, September 1985.
16. J. Waicukauski and E. Lindbloom, "Failure diagnosis of structured VLSI," *IEEE Design and Test of Comp.*, vol. 6, pp. 49–60, August 1989.

17. D.J. Burns, "Locating high resistance shorts in CMOS circuits by analyzing supply current measurement vectors," *Int. Symp. for Testing and Failure Analysis*, pp. 231–237, 1989.
18. S. Chakravarty and P.J. Thadikaran, "Simulation and generation of I_{DDQ} tests for bridging and leakage faults," Tech. Rep. No. 92-16, Dept. of Computer Science, State University of New York, Buffalo, NY 14260.

Sreejit Chakravarty B.E. Electronics and Electrical Engineering (1980) from Birla Institute of Technology and Science, Pilani, India; M.S. Computer Science (1983) from The State University of New York, Stony Brook; and Ph.D. Computer Science (1986) from The State University of New York, Albany. During 1980–81 he was a Development Engineer at Tata Electric Cos R&D Labs., Bombay, India. From 1986 to 1992 he was an Assistant Professor of Computer Science at the State University of New York, Buffalo where he is currently an Associate Professor of Computer Science. His research interests include: Algorithmic Aspects of VLSI, Fault-Tolerant Architectures, Parallel Algorithms.

Mingshen Liu received an MS in Computer Science from State University of New York at Buffalo in June 1992, a Master of Engineering in Computer Science and Application from the Institute of Computing Technology, Academia Sinica, China in 1986, and a Bachelor degree in Computer Science and Engineering from Beijing University of Aeronautics and Astronautics in 1983. He worked on CAD software development and CAD database while he was with the Institute of Computing Technology in China. He is now Software Engineer with Performance Technology, Inc., Rochester, NY, USA.

JOURNAL OF ELECTRONIC TESTING: Theory and Applications, 3, 387-396 (1992)

Proportional BIC Sensor for Current Testing

JOSEP RIUS AND J. FIGUERAS
Departament d'Enginyeria Electrònica, Universitat Politècnica de Catalunya, Avda. Diagonal 647, Barcelona 08028, Spain

Received January 30, 1992. Revised August 17, 1992.
Editor: C.F. Hawkins

Abstract. A new design of a BIC sensor for current testing static CMOS circuits is proposed. It is based on a lateral BJT device which is easy to incorporate in any standard CMOS process. The design diverts a fraction of the i_{DD} current from the cell under test and a resistive component generates a voltage proportional to i_{DD}. Additional features are the possibility of continuous measure of i_{DD} and increased speed of this sensor compared with sensors based on the current integration principle. The design does not have substrate currents due to the parasitic vertical BJTs. Experimental work on the sensor is reported.

Keywords: Built-in integrated sensor, CMOS lateral BJT, current test, gate controlled BJT, I_{DDQ} measure.

1. Introduction

Increasing attention in the test community is being given to the I_{DDQ} test methodology for static CMOS VLSI circuits. CMOS defects, such as bridges (resistive), gate oxide shorts and opens producing floating transistor gates are, in general, current testable and considerable research effort has supported this test methodology [1]–[8].

The current in a static CMOS cell (i_{DD}) is not constant through time. When an output or clock transition occurs, a peak of i_{DD} current is observed. This peak is due to charge and discharge of the load capacitance at the output circuit nodes and to overlap current through the PMOS and NMOS transistors in the circuit portion changing its state. When the transition is completed and the whole cell is in the quiescent state, i_{DD} (I_{DDQ}) is in practice near zero, and remains at this range until a new transition occurs. I_{DDQ} is very sensitive to circuit degradations or defects, which generate I_{DDQ} many orders of magnitude greater than the normal I_{DDQ}. This feature is used for the I_{DDQ} current test that detects defects.

The I_{DDQ} test through supply voltage pins of the IC package (the *external* current test) shows some problems because the major portion of of i_{DD} current in a static CMOS VLSI circuit is generated at the output pad circuits, and I_{DDQ} abnormalities are therefore shadowed by output currents. Moreover, large probe capacitances prolong test time and test frequency is low. The *internal* I_{DDQ} current test concept for static CMOS circuits has been proposed, based on Built-In Current sensor (BIC sensors) [5]. It is proposed as a solution for faster I_{DDQ} test rates.

The BIC sensor is built into the circuit being tested, and it can determine whether the circuit quiescent current is below or above a threshold level. Essentially, this technique adds a BIC sensor in the V_{DD} and/or GND lines to the block under test detecting abnormal I_{DDQ} levels in the block.

Characteristics required for the BIC sensor are:

- detection of abnormal static currents in a wide range of values.
- minimal disturbance of the static and dynamic characteristics of the cell under test.
- it be as simple and small as possible to minimize the additional area necessary to build it.
- the I_{DDQ} test should be fast.
- storing the fault state of the cell under test in order to read this state later.
- share as many circuits as possible for each BIC sensor in order to minimize the Si area requirements.

However, insertion of the sensor between V_{DD} and/or GND and the cell under test involves series voltages, and these voltages could degrade the function of the output cell. The BIC sensor terminal is either joined to all PMOS source electrodes (BIC sensor to

V_{DD}) or all NMOS source electrodes (BIC sensor to GND). This causes a large capacitance between these terminals and the substrate. This capacitance must be charged or discharged in a finite time and thus adds extra delays to the cell under test. The combination of both effects, a) effective lower supply due to series voltages and b) the additional capacitance delay time, might cause a degradation of the noise margin and/or speed of the cell under test. Designers must minimize these effects and achieve minimum disturbances in the circuit under test.

Also, the insertion of the BIC sensor may involve other consequences in the design of the circuits. For example, if the BIC sensor is connected to GND, then the sources of the NMOS transistors are disconnected from the GND (substrate in n-well technologies) forcing a review of certain layout design rules about distances, latch-up prevention, etc.

A method to reduce these undesired effects and degradations adds an extra pin to the IC. This pin must be joined to the point on the sensor that is connected to the circuit under test. For example, in BIC sensor to GND, this extra pin will be joined to all NMOS source electrodes. When testing is performed this pin is opened and the BIC sensor work is allowed. On the contrary, during normal operation this pin is grounded bypassing the BIC sensor. Thus, disturbances on the normal operation of the circuit can be avoided.

2. Proportional BIC Sensor

This sensor uses a built-in lateral BJT transistor (LBJT) in its structure that is connected in a common base configuration (Figure 1). This allows a fraction of i_{DD} current from the cell under test to be diverted to a resistive circuit that converts this current into a measurable voltage V_{sens}.

All i_{DD} current injected by the CMOS cell can pass through VBJT1 (vertical BJT) or LBJT and the division between both currents is constant and proportional to their respective emitter areas. Thus, to obtain a maximal efficiency lateral BJT and minimize the adverse effects of the VBJT1 some design rules must be followed [9], [10].

The insertion of the base-emitter junction between the source terminal from the NMOS transistors and V_{ss} is advantageous, with respect to the insertion of resistive elements, because the maximum voltage at the *VGND* node is limited by the forward biased junction in a wide range of i_{DD}. In addition, the negative

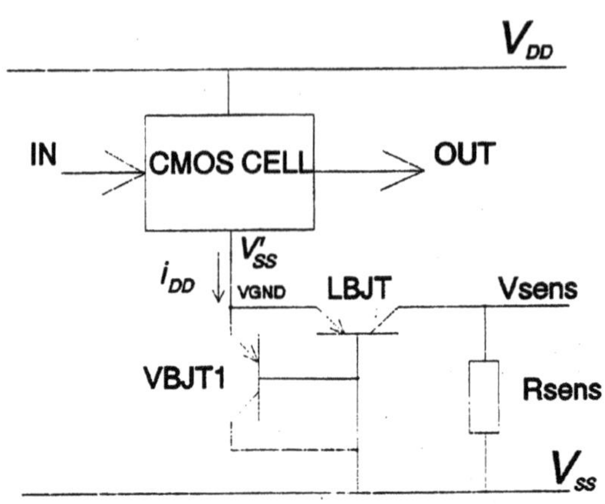

Fig. 1. Simplified schematic of the proportional sensor.

effect on sensor speed due to large capacitance in the *VGND* node is greatly diminished because of the exponential relation between lateral BJT base-emitter voltage and collector current. Thus, the collector voltage (V_{sens}) changes can be fast in response to base-emitter voltage changes, if the capacitance in the V_{sens} node is small. On the other hand, the proposed BIC sensor allows us to measure i_{DD} in a continuous fashion. An external circuit can decide when the measure will be made. For example, for I_{DDQ} test purposes, the best time can be immediately before a clock or input transition applied on the cell under test, because it guarantees that i_{DD} is in its quiescent value.

Our proposed sensor is very simple. Only one structure (that includes lateral and vertical BJTs), and a resistive component are necessary to get a voltage that reflects i_{DD}.

The test structure generates a voltage in the *VGND* node that must be subtracted from the cell output voltage. When the output transition finishes, a residual voltage V'_{ss} remains due to capacitance associated with this node. Thus, the logic *low* level at the cell output is not zero volts but the voltage V'_{ss}. However, this voltage remains at low values (around 0.6 V) if appropriate sizing of the lateral BJT is done. Moreover, the V'_{ss} voltage is not constant but varying between a maximum value in the output transition time and a minimum value. This minimum value depends on: a) the capacitance associated with the *VGND* node, b) the lateral transistor sizing and c) the frequency of cell switching. The result is a degradation in noise margin and speed of the circuit whose importance depends on the technology used and circuit characteristics.

2.1. *From V_{sens} to FAULT/NO FAULT Signal*

Additional circuitry must be added to the basic proportional BIC sensor circuit to extract a digital FAULT/NO FAULT signal from the analog V_{sens} signal comparing it with a voltage reference. The comparator circuit must be designed matching V_{sens} signal characteristics. Its main requirements would be:

- low voltage input levels: from a few mV to 800 mV approximately.
- low input offset.
- fast response, allowing high I_{DDQ} test frequency.
- simple, in order to minimize Si area.
- single power supply.

Comparators based on the sense amplifiers used in memory circuits fulfill these requirements.

2.2. *Lateral BJT in the Proportional BIC Sensor*

BJT structures using a pure CMOS process are well known [9]–[12] and it is possible to build lateral npn or pnp transistors depending on which process is used. A pair of vertical parasitic BJT can also be built into a lateral BJT structure. This lateral BJT structure is formed between the p^+ or n^+ diffusions and the well, while the parasitic vertical BJT structures due to interaction between the p^+ or n^+ diffusions, the well and the substrate.

Figure 2 shows two possible layouts that illustrate how the lateral BJT is made. Generally, layout (b) allows better performance (narrow base width) of the lateral BJT than layout (a) because the polysilicon width is usually less than the p^+ to p^+ diffusion distance. Transistors with layout (b) have been successfully implemented [11], [12], [13].

The gate terminal in Figure 2(b) must be properly biased to ensure bipolar behavior. This feature is achieved by biasing the adjacent MOS structure under the gate to the accumulation region, that is to say, by biasing the gate at a voltage with a polarity opposed to the majority charges sign on the well where the BJT is built. Thus, in n-well technologies, V_G will be a positive voltage.

SPICE models of the MOSFET transistors do not consider these bipolar structures formed by means of standard CMOS process. Therefore, it is not possible to simulate their behavior directly. The only way is to simulate these structures using their BJT equivalent model and getting the SPICE parameters of these equivalent BJTs from experimental measurements in lateral BJTs made with the standard CMOS manufacturing process.

These measures were performed on BJT structures built in the ECDM20 (n-well) CMOS process from ES2 [14]. The extracted parameters give coherent results and show clean bipolar work of such structures, hence allowing simulation of the proposed BIC sensor circuit.

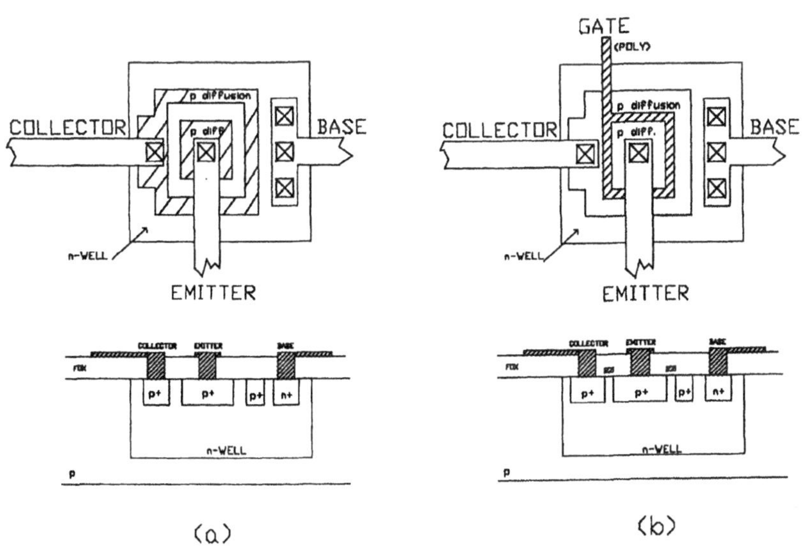

Fig. 2. Lateral BJT in CMOS n-well.

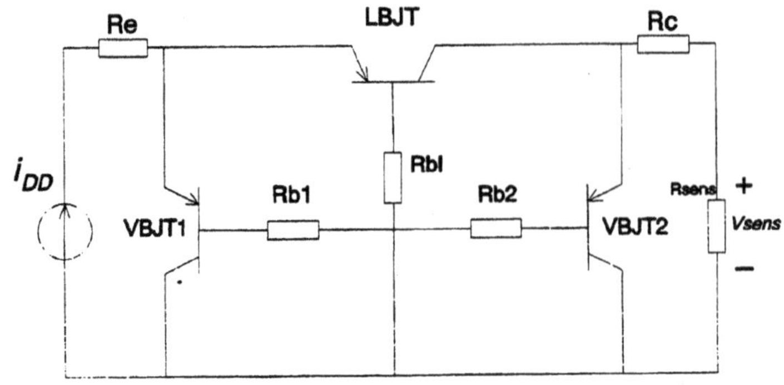

Fig. 3. Equivalent circuit of proportional BIC sensor.

3. Analysis and Simulation of the Proposed BIC Sensor

The transfer function between the injected i_{DD} and V_{sens} is necessary to derive the equivalent electric circuit and the sensor behavior equations. The equivalent electric circuit is shown in Figure 3. An analytical development was done [14] for the following simplifications to the full circuit (Figure 3).

a) Assume that the base resistances and base currents are small. Hence, the base voltage is near zero. Therefore, the actual VBJT1 and VBJT2 are diodes with their cathodes at GND.
b) Re is small and can be neglected.
c) Rc is included in the load resistance Rsens.
d) The actual diode VBJT2 will never be in conduction, because the output voltage V_{sens} will be smaller than its threshold voltage.
e) A DC analysis will be carried out using the Ebers-Moll model. The parameters of such a model will be considered as constants.

These five assumptions are valid for low i_{DD} values and time constants smaller than the cell switching period.

From the circuit equations we obtain the transfer function in implicit form:

$$V_{sens} = K\, i_{DD} - f(e^{V_{sens}/V_T} - 1) \qquad (1)$$

The interpretation of (1) can be made as follows:

For low i_{DD}, VBJT1 is off and LBJT is in the linear region. Thus, the first factor on the right hand of (1) is dominant, and V_{sens} *varies proportionally to* i_{DD}, that is, i_C is an i_{DD} fraction, whose magnitude (K) depends on transistor gain and lateral and vertical BJT saturation currents. In other words, it depends on the transistor sizes. A V_{sens}-i_{DD} graph in this zone follows a straight line. For this reason, we call this sensor a proportional BIC sensor. When the i_{DD} current increases, the LBJT transistor passes to the saturation region and VBJT2 starts to switch to the on-state. Thus, the second factor on the right hand of (1) is dominant and V_{sens} does not rise as fast as i_{DD}. If we operate the sensor outside the proportional range owing to a large rise of i_{DD}, V_{sens} will not increase proportionally and will maintain its value around 0.7 V due to the clamping of VBJT2.

Figure 4 shows a graph i_C(LBJT) = f(i_{DD}) with SPICE simulation using Ebers-Moll parameters. These were obtained from the measures performed on BJT structures built in the ECDM20 (n-well) CMOS process from ES2 indicated above; the figure also shows the calculated values from equation (1). As we can see, there is good agreement between simulated and calculated values in the proportional region (low i_{DD}).

4. Physical Layout Design of the Proposed BIC Sensor

In order to get preliminary experimental results about the proportional BIC sensor, some circuits with such structures have been designed and fabricated with ES2's ECDM20 (n-well) design rules. A simple lateral BJT was used: a standard lateral BJT made in CMOS process (Figure 5)—two rectangular p-diffusions (collector and emitter) separated by a thin layer of n-well material (base) below a poly gate. The transistor is surrounded by two rings. The inner ring is to bias the n-well (GND) and the outer is to bias the substrate (also GND) in order to avoid undesired currents.

The resistive circuit (Rsens) connected to the collector terminal of the lateral BJT was made with an

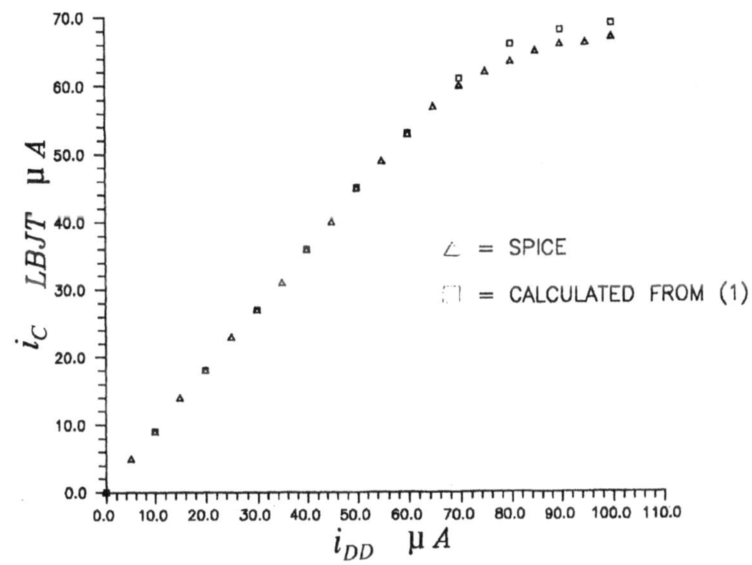

Fig. 4. $i_C = f(i_{DD})$ from (1).

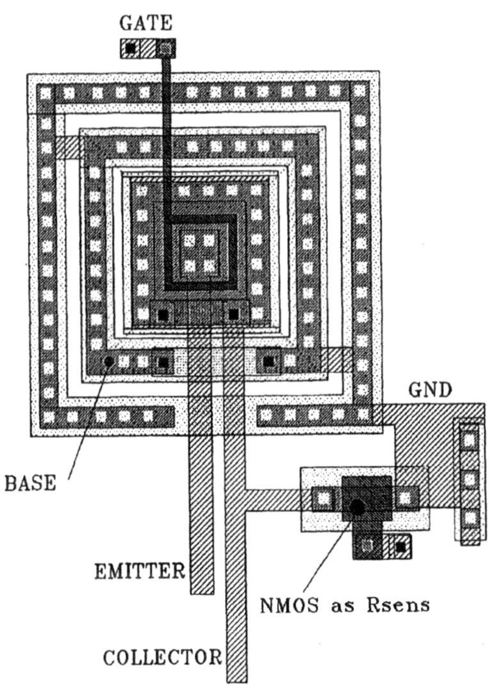

Fig. 5. Layout of lateral BJT shows in Figure 2(b).

NMOS transistor. Its gate was connected to a positive constant value (V_{DD}) in order to bias the transistor in the nonsaturated region and obtain its behavior as a resistor. Its size was adjusted to get an approximate value for its resistance of 11.5 kΩ.

Three circuits were made (Figure 6). In this figure the full schematic of the sensor (Figure 3) is simplified showing only LBJT, Rsens and the gate terminal of the BJT structure.

Circuit A includes an externally voltage controlled current source (VCCS) based on a current mirror applied to the BJT emitter. The collector electrode of such a transistor is connected to an NMOS transistor working as a resistor. The collector voltage (V_{sens}) is externally measurable. This circuit allows measurements of the transfer function $V_{sens} = f(I_{DD})$. Circuit B is the same as A but the gate voltage V_G of BJT structures is externally controlled, and measurement of the gate voltage influence on BJT behavior is therefore allowed. Circuit C is formed by a 2-NAND gate and a sensor. This circuit allows measurements of the sensor behavior when applied to a simple digital cell.

5. Experimental Measures on the Test Circuits

The circuits mentioned above were manufactured in ES2 and were tested in order to get the experimental transfer function and general behavior: static tests on circuit A, B, and C were made to verify the transfer

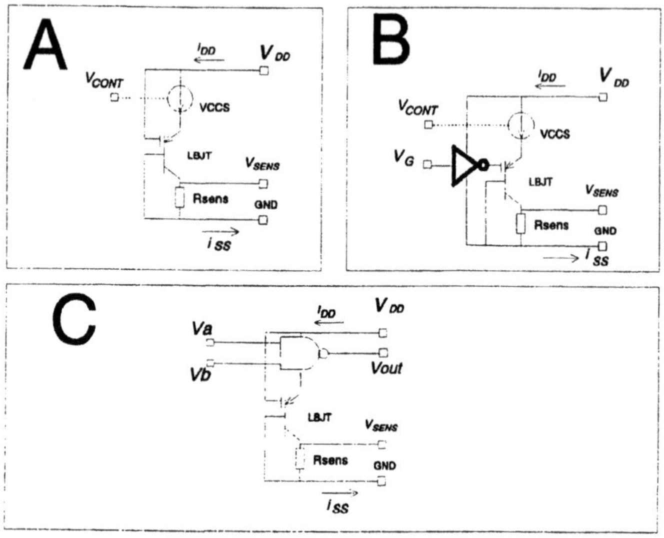

Fig. 6. Electrical diagram of circuits A, B, and C.

function between V_{sens} and i_{DD}. A static test on circuit B was made to verify the $V_{sens} = f(i_{DD}, V_G)$ characteristics.

5.1. $V_{sens} = f(i_{DD})$ Measures

A scan on V_{CONT} was made in circuit A, measuring simultaneously i_{SS} and V_{sens}. The results are shown in Figure 7. We can see the *almost proportional* rising of V_{sens} when i_{SS} passes from zero to 90 μA approximately. The experimental measurements shows the same V_{sens} qualitative dependence on i_{DD} as predicted in equation (1). However, some differences in the proportional region slope and the saturation voltage are observed. Figure 8 shows experimental and simulated transfer function curves. The simulated curve as obtained by using the experimental values mentioned above and sizing the BJT properly, Two simulated curves are shown. The first one has an Rbase = 850 Ω and the second one has Rbase = 0 Ω. These values were chosen as the minimum (ideal) value and a plausible value for the base resistance in order to show the strong dependency of the slope in the saturation region. The slope of the experimental curve in the saturation region lies between the slopes of the simulated curves.

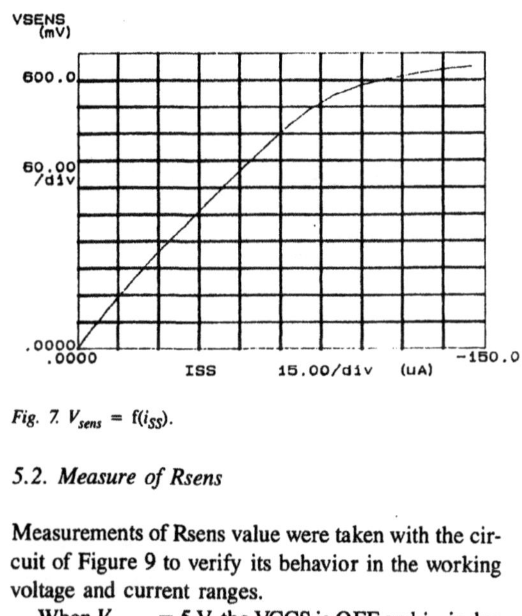

Fig. 7. $V_{sens} = f(i_{SS})$.

5.2. *Measure of Rsens*

Measurements of Rsens value were taken with the circuit of Figure 9 to verify its behavior in the working voltage and current ranges.

When $V_{CONT} = 5$ V, the VCCS is OFF and i_{SS} is due only to i_{sens} injected. The injected current circulates through Rsens and generates V_{sens}. While $V_{sens} < 0.6$ V the current injected in the base-collector junction may be neglected. Figure 10 shows the Rsens = V_{sens}/i_{sens} curve labeled as (1) and $i_{SS} = i_{sens}$ curve labeled as (2).

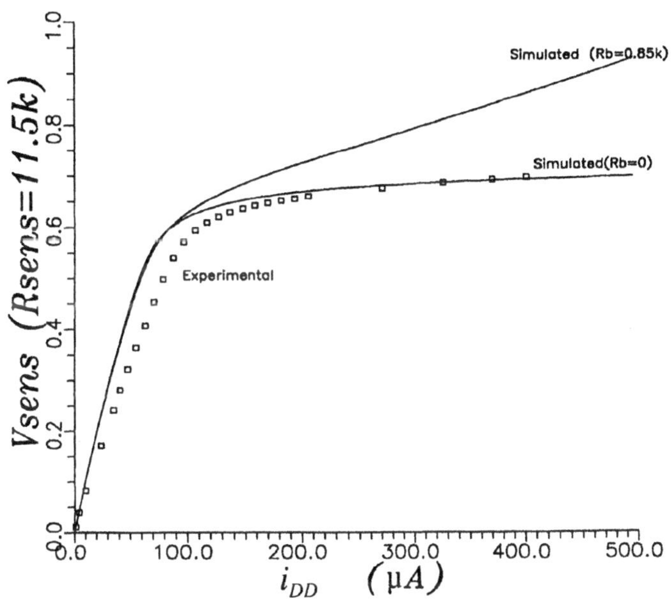

Fig. 8. Experimental and simulated results on transfer function on circuit A.

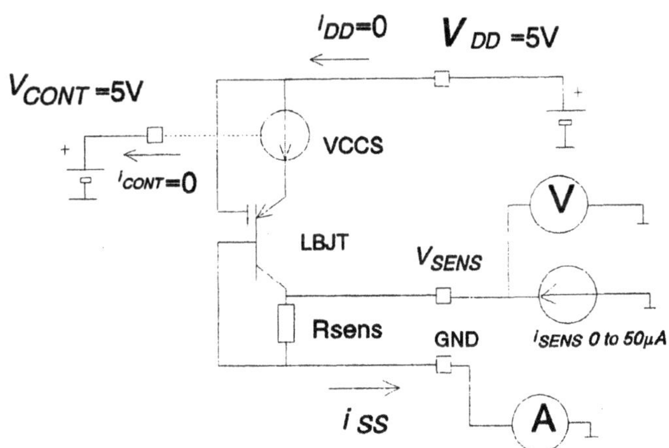

Fig. 9. Circuit for measuring Rsens.

5.3. Measure of Voltage Gate Influence on $V_{sens} = f(i_{DD})$

A test was made on circuit B using an HP4145B curve tracer and the circuit in Figure 11.

By varying V_{CONT} between 0 and 5 V and taking V_G as a parameter, it is possible to obtain the influence of V_G on the transfer function. In Figure 12 we can see $V_{sens} = f(i_C, V_G)$ instead of $V_{sens} = f(i_{SS}, V_G)$ but i_C is equal to the current through VCCS, therefore, Figure 12 is equivalent to Figure 7.

The measures show a small change in the proportional BIC sensor performance when the gate voltage changes from 0 V to 5 V. On the other hand, if the base voltage equals zero volts (n-well connected to V_{SS}) and if the holding gate voltage is constant, then the sensor behavior is stable and predictable.

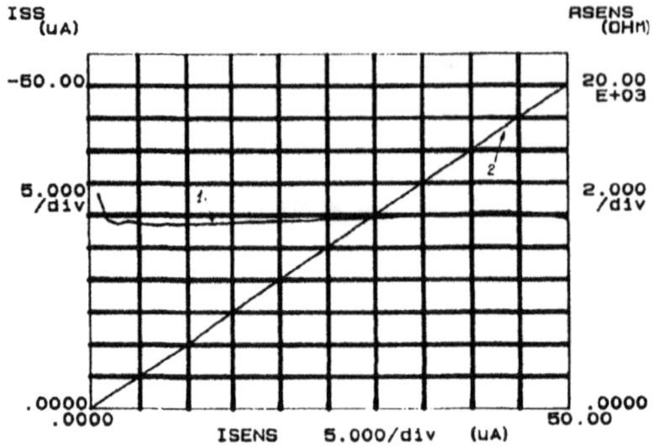

Fig. 10. Rsens dependence on V_{sens}. ($1 \rightarrow$ Rsens $= V_{sens}/i_{sens}$, $2 \rightarrow i_{SS} = i_{sens}$).

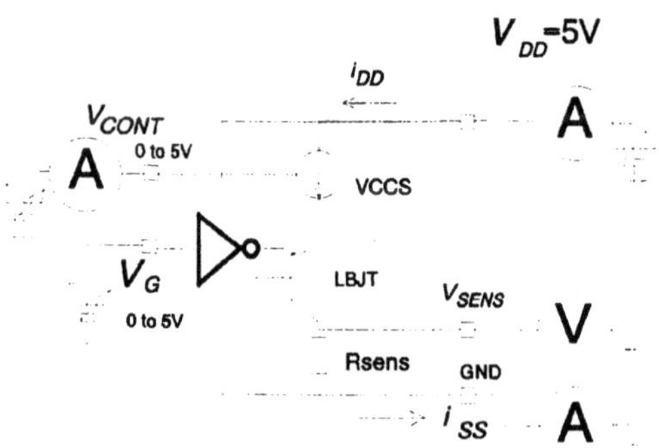

Fig. 11.. Circuit for measuring the gate voltage influence on circuit B.

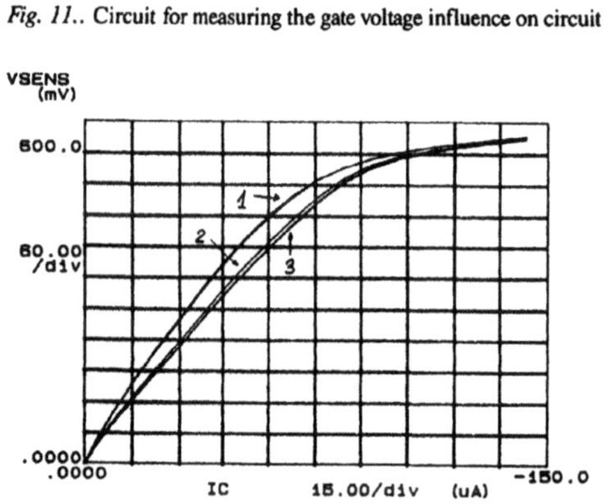

Fig. 12. Dependence of V_{sens} = f(i_C, V_G). $1 \rightarrow V_G = 4$ and 5 V, $2 \rightarrow V_G = 3$ V, $3 \rightarrow V_G = 0$, 1 and 2 V.

5.4. Sensor Applied to a 2-NAND Gate. Static Measures

The test was made on circuit C (Figure 13) using an HP4145B curve tracer.

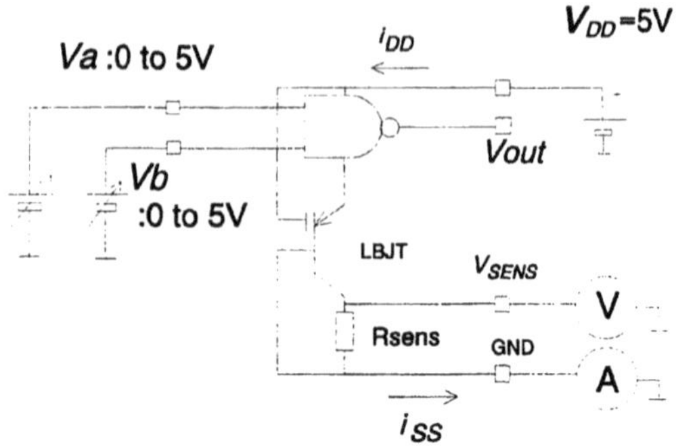

Fig. 13. Circuit C for static measurements on a NAND circuit with a proportional BIC sensor.

Vb was set at fixed voltages from 0 V to 5 V while i_{SS} and V_{sens} were measured as a function of Va (Figure 14). The experimental measures (Figure 14) agree with the simulations (Figures 15 and 16) and point out the qualitative feasibility of the proposal.

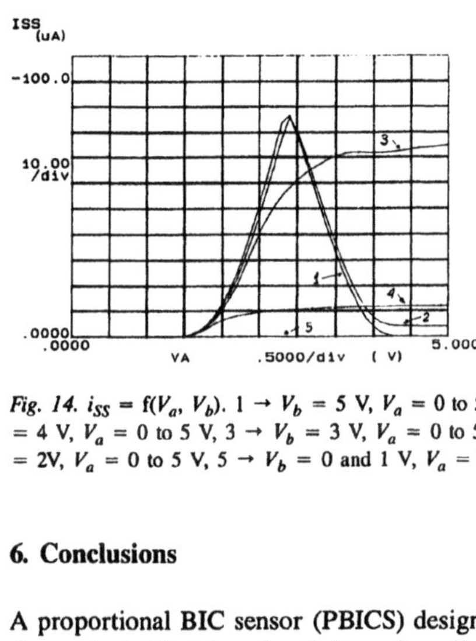

Fig. 14. i_{SS} = f(V_a, V_b). $1 \rightarrow V_b = 5$ V, $V_a = 0$ to 5 V, $2 \rightarrow V_b = 4$ V, $V_a = 0$ to 5 V, $3 \rightarrow V_b = 3$ V, $V_a = 0$ to 5 V, $4 \rightarrow V_b = 2$V, $V_a = 0$ to 5 V, $5 \rightarrow V_b = 0$ and 1 V, $V_a = 0$ to 5 V.

6. Conclusions

A proportional BIC sensor (PBICS) design has been demonstrated that thas simplicity and small size. It can supply a voltage proportional to the current injected in a wide range of values and allow continuous measure of this current. Its I-V characteristic limits its drop voltage for a wide range of i_{DD} values to approximately 0.7 V. Also, the exponential dependence of the V_{sens} voltage versus the voltage of the *VGND* node allows a fast response. The design avoids the large substrate currents due to the parasitic vertical BJT with its collectors joined to the substrate. In the design the residual gate effect on lateral BJT behavior can be neglected

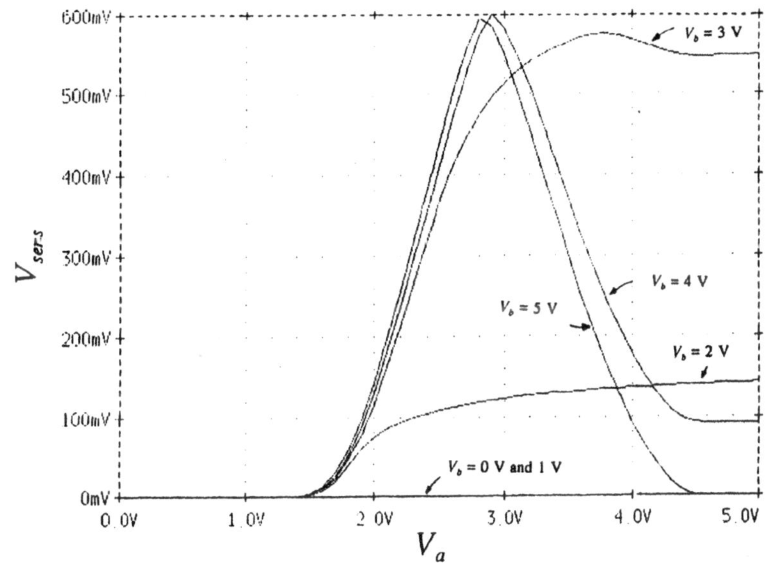

Fig. 15. Simulation of a PBICS applied on a 2-NAND V_{sens} = f(V_a, V_b).

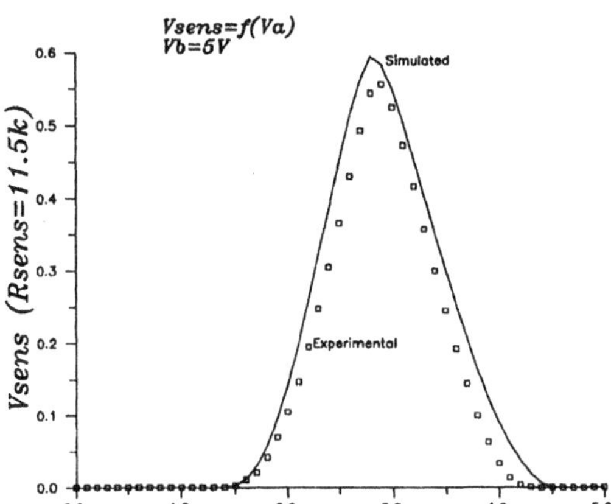

Fig. 16. Experimental and simulated values on V_{sens} = f(V_a, V_b.

because of the voltage applied between the LBJT gate and the substrate is constant. The unavoidable noise margin and speed degradation can be held to the minimum by means of the appropriate sensor sizing. The experimental results of the proposed BIC sensor show a good agreement with the values predicted.

References

1. C.F. Hawkins and J.M. Soden, "Electrical characteristics and testing considerations for gate oxide shorts in CMOS ICs," *Proc. Int. Test Conf.*, pp. 544–555, November 1985.
2. C.F. Hawkins and J.M. Soden, "Reliability and electrical properties of gate-oxide shorts in CMOS ICs," *Proc. Int. Test Conf.*, pp. 443–451, September 1986.
3. L.K. Horning, J.M. Soden, R.R. Fritzemeier, and Ch.F. Hawkins, "Measurements of quiescent power supply current for CMOS ICs in production testing," *Proc. Int. Test Conf.*, pp. 300–309, September 1987.
4. M. Jacomino, J.-L. Rainard, and R. David, "Fault detection in CMOS circuits by consumption measurements," *IEEE Trans. on Instrumentation and Measurement*, vol. 38, pp. 773–778, June 1989.
5. W. Maly, P. Nigh, P.K. Nag, D.B.I. Feltham, and L.R. Carley, "Built-in current testing. Part I," *Research Report No. CMUCAD-88-27*, June 1988.
6. Y.K. Malaiya and S.Y.H. Su, "A new fault model and testing technique for CMOS devices," *Proc. Int. Test Conf.*, pp. 25–34, 1982.
7. R. Rodriguez, J. Segura, V. Champac, J. Figueras, and A. Rubio, "Current versus logic testing of bridges, GOS and floating gates in CMOS circuits," *Proc. Int. Test Conf.*, Nashville, TN, October 1991.
8. A. Rubio, J. Figueras, and J. Segura, "Quiescent current sensor circuits in digital VLSI CMOS testing," *Electronic Letters*, vol. 26, pp. 1204–1206, July 1990.

9. E.A. Vittoz, "MOS transistors operated in the lateral bipolar mode and their application in CMOS technology," *IEEE Journal of S.S. Circuits*, vol. SC-18, pp. 273–279, June 1984.
10. M.R. Haskard and I.C. May, *Analog VLSI Design, NMOS and CMOS*, Sydney: Prentice Hall, 1988.
11. M.G. Degrauwe, O.N. Leuthold, E.A. Vittoz, H.J. Oguey, and A. Descombes, "CMOS voltage references using lateral bipolar transistors," *IEEE Journal of S.S. Circuits*, vol. SC-20, pp. 1151–1157, December 1985.
12. T.W. Pan and A.A. Abidi, "A 50 dB variable gain amplifier using parasitic bipolar transistors in CMOS," *IEEE Journal of S.S. Circuits*, vol. 24, pp. 951–961, August 1989.
13. X. Arreguit and E.A. Vittoz, "Performance enhancement of compatible lateral bipolar transistors for high precision CMOS analog design," *Proc. ESSCIRC'88*, Manchester UK, September 1988.
14. J. Rius, "Analisi i propostes de millora de sensors de corrent integrats. Disseny d'un sensor per test de corrent en C.I. CMOS," Final Project Report No. UPCDEE-91-2, February 1991.

Josep Rius received engineering degree from the Universitat Politècnica de Catalunya (ETSEIB). He is associated professor at the Electronics Engineering Department of the Politechnical University of Catalonia (UPC) since 1991. His current research interests include current characterization and testing techniques of digital cells.

Joan Figueras received his Industrial Engineering (Electrical) degree from the Universitat Politècnica de Catalunya (U.P.C.) and the Ms.C. (Electrical) and Ph.D. (Computer Information and Control) degrees from the University of Michigan where he was a Fulbright Scholar and worked in the Systems Engineering Laboratory (1968–71). He has been visiting scholar in UCLA (1979) and UCI (1992). Since 1983 is Professor at the Electronics Dept. of the U.P.C. with teaching and research responsibilities in the area of Digital Systems and VLSI design. His current research interests include Test and Design for Testability, Defect and Fault Models and new Testing and Fault Tolerance methodologies.

JOURNAL OF ELECTRONIC TESTING: Theory and Applications, 3, 397-406 (1992)

Design of ICs Applying Built-In Current Testing*

WOJCIECH MALY AND MAREK PATYRA
Electrical and Computer Engineering Department, Carnegie Mellon University, Pittsburgh, PA 15213

Received February 5, 1992. Revised August 24, 1992.
Editor: C.F. Hawkins

Abstract. Built-in Current (BIC) sensors have proven to be very useful in testing static CMOS ICs. In a number of experimental ICs BIC sensors were able to detect small abnormal I_{DDQ} currents. This paper discusses the design of the circuit under test and Built-in Current (BIC) sensors, which provide: maximum level of defect detectability, minimum impact of BIC sensor on the performance of the circuit under test and minimum area overhead needed for BIC sensors implementation.

Keywords: Built-in testing, BIC-testing, current testing.

1. Introduction

I_{DDQ} testing [1]-[4] was found to be very useful in detecting large classes of faults occurring in tested ICs [5], [6]. Many researchers, as well as years of experience, have shown that this testing technique, especially useful in static CMOS circuits, can provide a quality of testing unattainable by any other testing methods [4], [7]. It was also recognized that implementation of I_{DDQ} testing has not been straightforward because it has required both special purpose testing hardware and a significant decrease in the testing rate [7]. To overcome these shortcomings a Built-in Current (BIC) testing technique was proposed [7], [8] and a number of experimental IC chips with BIC sensors were fabricated and tested [9]-[11]. The BIC testing implemented on these chips applies current sensors to monitor the amount of quiescent current in the power lines of the functional units under the test. Since the vast majority of defects occurring in the typical process causes an increase in the quiescent current [5], [7], BIC sensors can be used effectively for testing purposes.

Fabrication experiments conducted in the last couple of years, have fully confirmed the basic concept of BIC testing. They also helped to answer three key questions: (1) how to minimize the impact of the BIC sensor on the performance of the functional unit under test; (2) how to optimize the BIC sensor design from the defect detectability point of view; and (3) how to optimize the circuit under test to obtain the maximum level of defect detectability at the same time minimizing area overhead needed for BIC sensors. The goal of this article which is an extended version of [5] is to summarize the information addressed by the above questions.

*This research was supported by NSF Grant MIP8822805.

2. BIC Testing

The basic idea of BIC testing is simple and has been explained in a number of articles [5], [7], [8], [12]. This section summarizes the key elements of the BIC testing methodology published in the past and also discusses a newer version of the BIC sensor design.

2.1. *Implementation of BIC Testing*

Figure 1a explains the basic structure of the BIC sensor. It is composed of a voltage drop device and a voltage comparator. These basic components of any BIC sensor are arranged such that at the end of each clock cycle the virtual ground voltage, V_{GND}, is compared with the voltage reference, V_{ref}. The value of V_{ref} is chosen such that $V_{GND} < V_{ref}$ for defect free functional units and $V_{GND} > V_{ref}$ for those units where the quiescent current indicates an occurrence of a defect. It is also possible to locate the voltage drop device

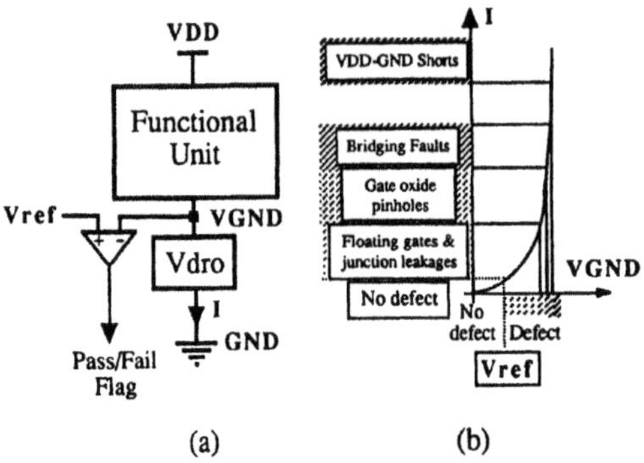

Fig. 1. BIC sensor.

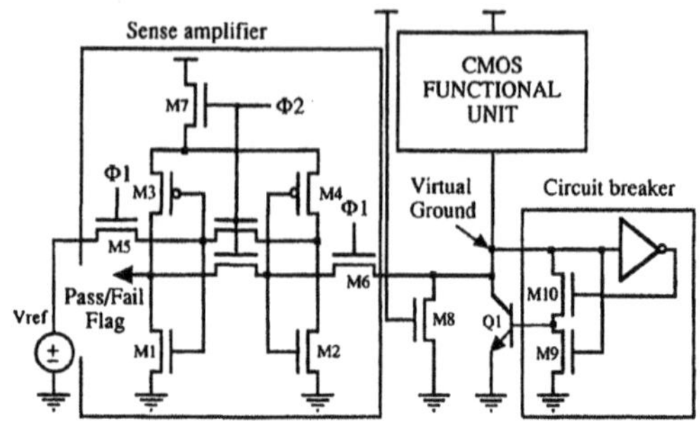

Fig. 2. BIC sensor implemented in the CMOS technology.

between the circuit under the test and V_{DD} and appropriately adjusting V_{ref}.

The voltage drop device impacts the operation of both the BIC sensor and funtional unit under the test. The problem is the large magnitude of currents which must be handled by the voltage drop device. On one hand, the voltage drops across this device—which are due to the normal peak currents—should be small. Voltage drops generated by such defects as *VDD-GND* shorts, and various bridging should not be too large either. On the other hand voltage drops which are due to very small, defect-generated currents, such as current caused by gate oxide pinholes, should be large enough to be measurable by the voltage comparator. Consequently, voltage drop devices having linear $I(V)$ characteristic (resistor) can be effective either for detection of defects producing small abnormal currents or can have a value minimizing its impact on normal operation of the functional unit under the test. One cannot achieve both unless a nonlinear element, having a characteristic such as one shown in Figure 1.b, is used. Therefore, a bipolar device rather than a MOS device should be used as a voltage drop element if a high resolution of current detection is required.

The idea depicted in Figure 1 can be implemented in CMOS technology in a variety of ways. Figure 2 shows a CMOS implementation of a BIC sensor [8], [9] which employs a substrate bipolar npn transistor as a voltage drop device Q1. This version of the BIC sensor was designed for self testing and reconfiguration purposes [13]. It has a simple sense amplifier and a circuit breaker which can be used to disconnect a defective functional unit from the power rail in case of large abnormal currents caused by *VDD-GND* shorts or radiation generated latch-up. (Transistor M8 is used to sink very low current from the disconnected functional unit.). This version of BIC sensor was used, for instance, in a self-testing ALU described in [14]. The application of this design was, however, not trivial since it requires two clocks (Φ_1, Φ_2) and reference voltage V_{ref}, which was generated off-chip.

Figure 3 depicts another BIC sensor implementation intended for a fault tolerant system. The design is based on the idea proposed in [10]. The BIC sensor shown in Figure 3 uses: a voltage drop device (transistor Q1), a comparator (M1-M7), a two-stage amplifier (M3, M8, I1), a bistable edge-sensitive latch (LATCH) generating flag Pass/Fail flag, a circuit breaker (M9, M10, I2), a reference voltage source (*Vref*) and a current source. The circuit sets the Pass/Fail Flag at "1" and disconnects the CMOS functional unit from ground when V_{CE} of Q1 is higher than V_{ref}, and *Control* signal is high. The timing of control signals is relatively simple. Within the system clock cycle the BIC Resets (RES) falling edge sets the Pass/Fail Flag at "0." The BIC clock's (CKL) rising edge latches the data from the comparator and sets the Pass/Fail Flag at "1" in case of defect occurrence (the Control signal must be high at that moment).

The essential elements of the above BIC sensor implementation are: the application of a bipolar transistor; the low offset comparator, and an adequate V_{ref} voltage source. Our experience indicates [8]–[11], that the exponential characteristic of the voltage drop device (see Figure 1b) is very important in minimizing the impact of the BIC sensor on the performance of the unit under test especially for high magnitude switching transient currents. The exponential characteristic also helps to achieve high resolution of low abnormal I_{DDQ} currents.

Of course, in the case when application of bipolar device is not desirable one can also use a MOS transistor as the sensor. Figure 4 shows a design of a fast BIC sensor with a standard MOS transistor (labeled

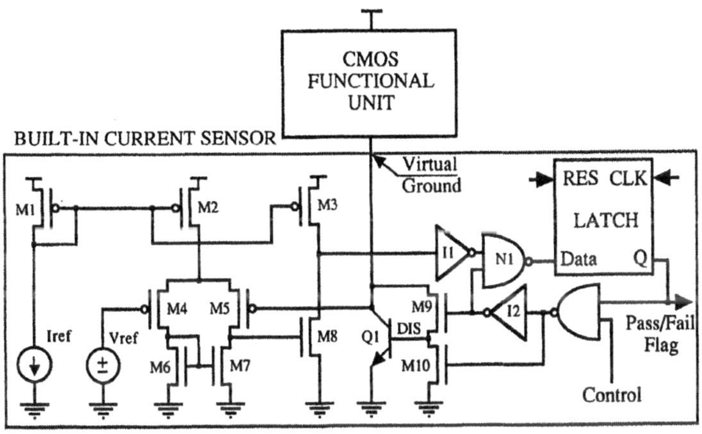

Fig. 3. BIC sensor for application in a fault tolerant IC.

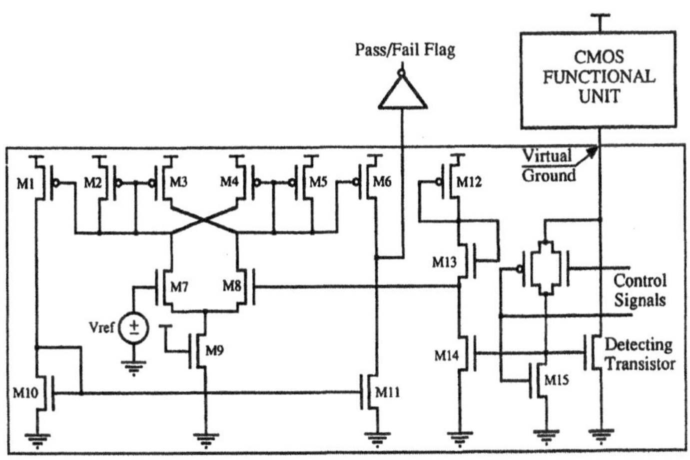

Fig. 4. BIC sensor using MOS transistor as a voltage drop device.

detecting transistor) used as the voltage drop element. It is also possible to build sensors with any combination of bipolar transistors, diodes, and MOS transistors which can provide low voltage drop for peak current and adequate voltage drop for low abnormal I_{DDQ} currents. Later in this article we assume that the applied BIC sensors have all the positive attributes of the sensors described in this section and we do not again discuss the design of the sensor itself.

2.2. *Possible Application of BIC Sensors*

Essentially, BIC testing can be applied in two modes of operation: in *die selection mode* and *concurrent passive testing mode*. In the first mode, BIC testing can detect permanent manufacturing defects during probe or package testing, in which testing vectors can be generated off-chip or on-chip. In the second mode, the BIC sensor can passively monitor currents during normal IC operation. When a fault is detected due to both permanent manufacturing defects or defects caused by reliability problems, then the BIC sensor can generate a fail flag that subsequently is used to perform the appropriate action. These outlined options can be used in various combinations. More discussion of the possible applications of BIC testing can be found in [11], [12], [14].

2.3. Design of the IC with BIC Sensors

Notice that regardless of the BIC sensor implementation, there are a number of general issues that must be addressed by the designer of the system applying such sensor. Three most important of them are related to the questions:

1. How to minimize the impact of a BIC sensor on the performance of the circuit under the test (CUT)?
2. How the CUT should be partitioned into functional units, each covered by a single BIC sensor?
3. How to design each unit of the system to achieve maximum utilization of the potential capabilities provided by the concept of BIC testing?

The goal of this article is to answer the above questions. But, as was mentioned previously, we will focus on the last two questions because the answer to the first has already been discussed in the other articles.

3. IC Partioning

Assume that a digital system to be implemented on a single IC must be partioned into functional units—each covered by a single BIC sensor. Regardless of the function performed by each unit and what is the specific functions performed by the system, one must decide:

1. What should be the objective function for the partioning procedure?
2. What is the optimal number of transistors in each partition?

To handle the above questions we will assume that the design objective is simply BIC sensor minimum area overhead. (This can be translated into the minimization of the number of BIC sensors in the IC). Consequently, one can solve the circuit partitioning problem if the maximum number of transistors in a circuit monitored by a single sensor is known. In this section we discuss these questions in more detail.

3.1. Size of a Functional Unit Covered by a Single BIC Sensor

Figure 5 illustrates the basic elements involved in the choice of the size of the functional unit covered by a single BIC sensor. These elements are:

1. I_{defmin}—the smallest current which must be identified by a sensor as a symptom of a defect.
2. I_{DDQ}—the quiescent current of the functional unit under the test (which at this stage will be assumed to be a linear function of the number of transistor in the circuit under test).
3. $I_{noisemax}$—the maximum noise related peak current.

The region in Figure 5 called *design space* represents a range of currents which can be utilized to indicate a defect free operation of the unit under the test. Notice that to distinguish between defect-free and defective unit under the test, it is necessary to build a functional unit such that the total quiescent current generated by the defect free unit under test is less than I_{defmin}. The solution which minimizes the area overhead used by a BIC sensor is, then, located at the intersection of I_{defmin} and I_{DDQ} (under the condition that $I_{defmin} > I_{noisemax}$). This intersection represents a maximum size unit that produces quiescent current equal to I_{defmin}.

Hence, in order to find N_{max} one must determine the actual value of the current through the ground (or V_{DD}) rail of the functional unit under test at the end of the clock cycle. Such current must be determined as a function of the number of transistors in a unit under the test. To find this current one must carefully analyze all of its components.

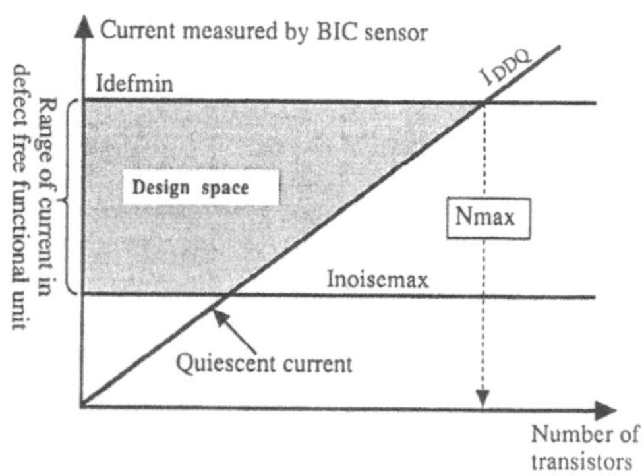

Fig. 5. Optimum number of transistors in a single unit.

3.2. *Leakage Currents in Defect Free Circuit*

To see basic components constituting I_{DDQ} let us consider the chain of two inverters. As shown in Figure 6, I_{DDQ} is essentially composed of junction leakages and subthreshold channel leakages. Junction leakages are of two types: state related (compare Figure 6a and b) and state independent (caused by well reverse bias).

3.2.1. Junction Leakages. The amount of reverse biased junction leakage currents in the static CMOS circuit can be determined by evaluating the area of all reverse-biased *p-n* junctions in the circuit, i.e., the well-substrate junction and approximately one half of all circuit's drain and source junctions. It is assumed that half of the static CMOS circuit's drains or sources are reverse biased. (See Figure 6.)

Hence, one can more precisely determine the drain/source related component of I_{DDQ} by:

Step 1: Computing the relationship between the drain leakage current and the drain area. Such a relationship is shown in Figure 7. Note that this figure shows $I_g(A)$ for a spectrum of saturation currents I_{gm}.

Step 2: Determining the actual value of I_{gm} using an appropriate test structure. This determines which family of curves is equivalent to the curves shown in Figure 7, and should be used.

Step 3: Computing the total current using the formula [15]:

$$I_g = I_{gm} \sum_{i=1}^{M} \left[W_i K_i + \pi Xj(W_i + K_i) + \frac{1}{6} \pi Xj^2 \right] \tag{1}$$

where: M is the number of drain regions; W_i is a width; K_i is a length of the i-th drain region; and X_j is a drain junction depth.

The same procedure must be applied to all well junctions but with a different set of curves (see example in Figure 8) and noting that all walls are reverse biased all of the time.

The well junction current can be found, for instance, using the formula:

$$I_{gw} = I_{gm} \sum_{k=1}^{N} \left[Y_k Z_k + 1.6Wj(Y_k + Z_k) + 0.8\pi Wj(Y_k + Z_k) + \frac{1}{6} \pi (0.8Wj)^2 \right] \tag{2}$$

where N is a number of well areas and Y is width, Z is length and Wj is well junction depth.

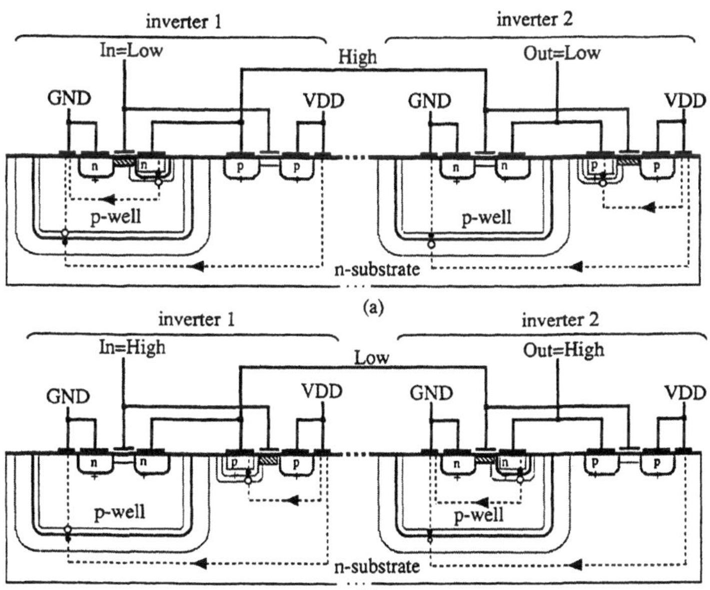

Fig. 6. The chain of CMOS inverters with In = LOW (a) and In = HIGH (b).

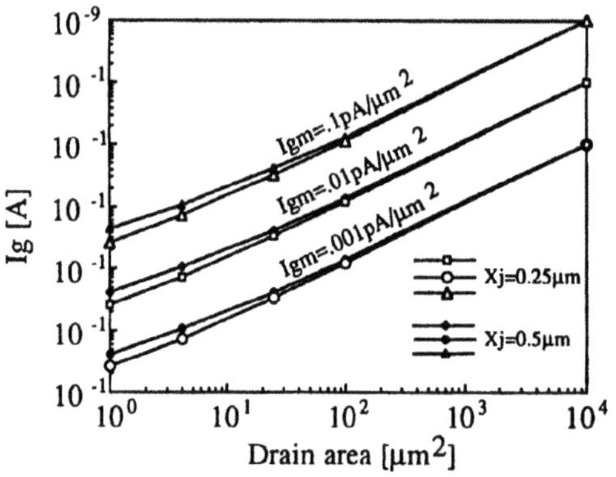

Fig. 7. Leakage current calculated as a function of the well area.

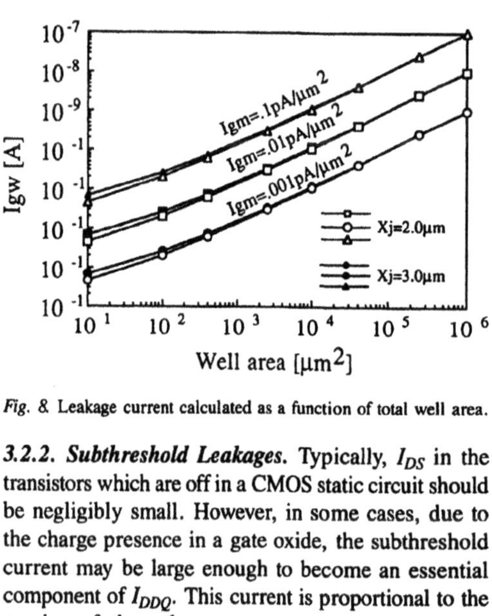

Fig. 8. Leakage current calculated as a function of total well area.

3.2.2. Subthreshold Leakages. Typically, I_{DS} in the transistors which are off in a CMOS static circuit should be negligibly small. However, in some cases, due to the charge presence in a gate oxide, the subthreshold current may be large enough to become an essential component of I_{DDQ}. This current is proportional to the number of channel squares.

3.3. Transient Currents in I_{DDQ} Measurements

In the discussion presented so far it was assumed that the BIC sensor should measure I_{DDQ} at the end of the clock cycle. Indirectly then, it was assumed that the clock cycle is long enough to allow the circuit to reach complete stability. In reality such an assumption is unacceptable for the reason illustrated in Figure 9. The problem is that the time needed for the current to decay is much longer than the time needed for the circuit's nodes to reach their correct logic states.

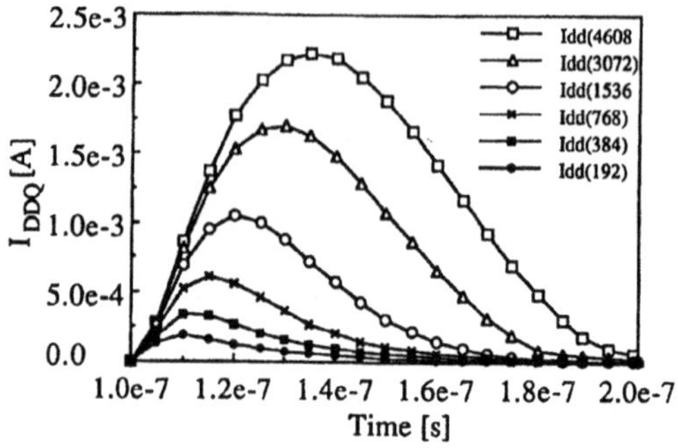

Fig. 9. Simulated I_{DDQ} current decay for different number of transistors in a CMOS circuit.

Consequently, the I_{DDQ} monitored at the end of the clock cycle—the duration of which is dictated by the circuit performance only—will be composed of the quiescent component discussed previously and a transient component of the decay current. In many cases the *decay component* may be much larger than the leakage component itself (notice that all *recharged capacitors* will *consume charge* even if all appropriate voltage value are already above logic thresholds). For instance, Figure 10 shows what the value of the decay current would be in a functional unit fabricated in standard 2 μm CMOS technology if the current measurements are done at the end of a normal clock cycle and assuming very low load values. These curves would look even worse if large load capacitors were assumed for the nodes connected to the output pins.

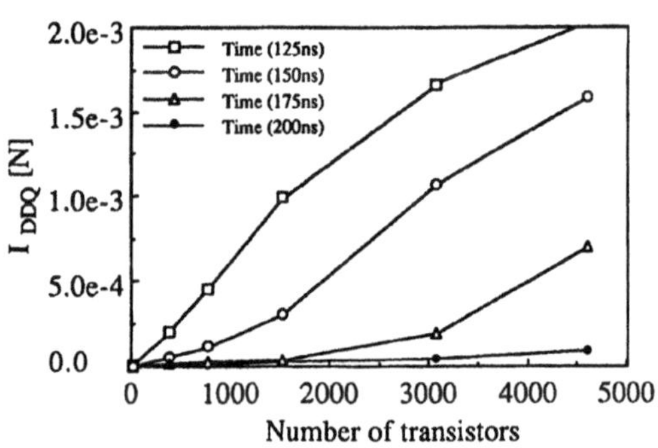

Fig. 10. Simulated I_{DDQ} quiescent current as a function of number of transistors for four different clock cycles.

3.4. Optimal Choice of N_{max}

Let us return to the diagram presented in Figure 5 to reexamine the strategy for choosing N_{max} the number of transistors in a functional unit covered by a single BIC sensor. As previously, we must assume that N_{max} should be such that I_{DDQ} (including its decay component), measured at the end of the given clock cycle, should be smaller than the minimum abnormal current caused by the actual defect decreased by the offset voltage of applied voltage comparator. Hence, the algorithm for choosing N_{max}, (Figure 11) should be as follows:

Step 1: A set of I_{DDQ} currents, expressed in terms of the size of the functional unit and the clock cycles, should be determined.[1]

Step 2: Choice of the minimum abnormal (defect related) current, decreased by the offset current representing the inherent offset of the voltage comparator, should be made.

Step 3: Choice of the clock period used during testing (of the normal clock cycle is not used) should be made and an equivalent I_{DDQ} curve should be determined.

Step 4: The intersection of I_{defmin}-I_{offset} and I_{DDQ} curves should be determined.

Notice, that a strong tradeoff exists between the clock rate used during testing and the BIC sensors' ability to cope with defects causing small abnormal I_{DDQ} currents. This indicates that the IC designer may consider a number of modes of operation for the IC under the test—a concurrent mode where the circuit is driven *full speed* and a testing mode where the circuit is *slowed down* for a number of clock cycles while the emerging low current defects are detected.

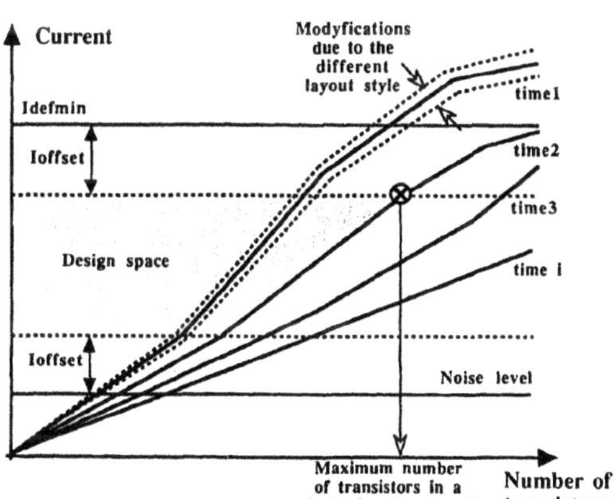

Fig. 11. Strategy for the choice of N_{max}.

4. Circuit Level Design

The efficiency of BIC testing is a function of the amount of normal quiescent current seen by the BIC sensor. Therefore, design of a circuit with BIC sensors should be performed in a manner that minimize this current. Designers should avoid circuit solutions different from standard static gates.[2] They also should be careful with charge sharing problems which may cause some devices in the circuit to operate in the subthreshold region, thereby causing abnormal levels of I_{DDQ}.

The other key issue related to the circuit level design of a system using BIC sensors relates to the fact that two different functional units connected to each other, but having two separate BIC sensors may interact, in a way causing large power supply leakages or, in the extreme case cause latch-up. In this section we will discuss the above issues in more detail.

4.1. Currents in the Interface between Units

Consider a typical situation which may occur in a CMOS circuit between two separate units connected to the global ground (*GND*) through two independent BIC sensors (Figure 12). Notice that in such a case the transient components of the virtual ground voltages, *VGNDi* and *VGNDj*, may be different when measured with respect to the global ground. Consequently, it is likely that for a period of time *VGNDj* > *VGNDi*. If we now assume that the pass gate between these two units (Figure 12) has an n-channel transistor placed in the well which is biased using *VGNDj*, then the BJT parasitic transistor, formed by the n-type source of the n-channel transistor (emitter), p-well (base) and n-type substrate (collector) is in the active mode of operation. Hence, for a period of time when *VGNDj* > *VGNDi*, a large current exists between *VGNDi* and *VDD*. This current may also trigger latch-up which will continue even after the period of time when *VGNDj* > *VGNDi*. There are a number of ways to prevent this phenomenon. The most effective seems to be a simple design rule that the output of any functional unit should see only the gate oxide in the unit covered by a separate BIC (Figure 13). In particular it should not be connected to the pass gate belonging to the other unit.

4.2. Layout Optimization

An important characteristic of the circuits for which BIC testing was invented is an unfavorable ratio of

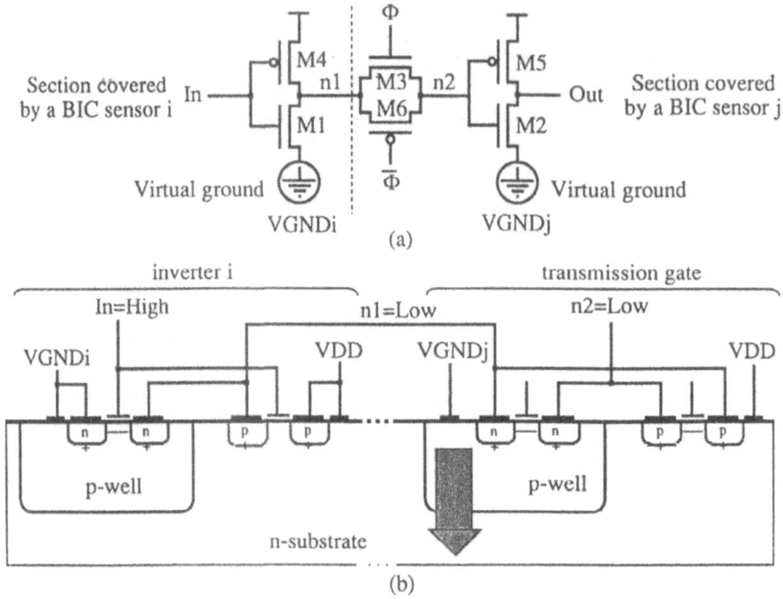

Fig. 12. An example of interface between units with separate BIC sensors.

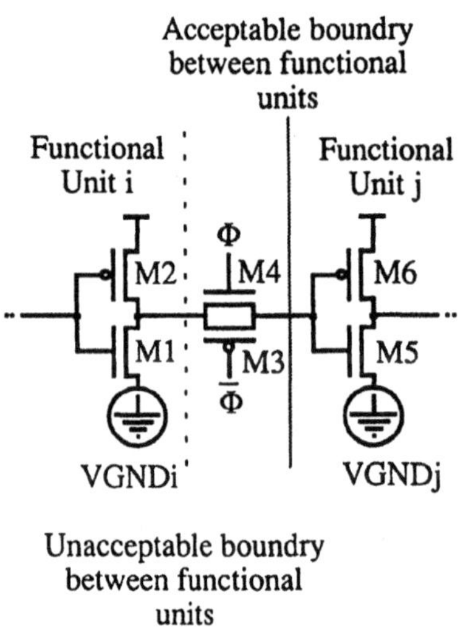

Fig. 13. Good and wrong location of the boundaries between functional units monitored by two different BIC sensors.

circuit delay time to the current decay time. Therefore, circuits which are intended to be monitored by different BIC sensors, should be designed in a manner maximizing this ratio.

To see that such a strategy is possible, consider a tri-state inverter shown in Figure 14. It can be implemented in a number of ways. Figure 15 shows five different CMOS layouts of this inverter. The straightforward design (Figure 15a) results in a very small parasitic capacitance associated with the output node and high parasitic capacitances on the internal nodes n1 and n2. This causes high delay and decay times. Therefore, in the subsequent implementation (Figure 15a–e) an effort was made to decrease capacitance on the internal nodes n1 and n2, even if value of C_{OUT} had to be increased. The extent of the improvement is

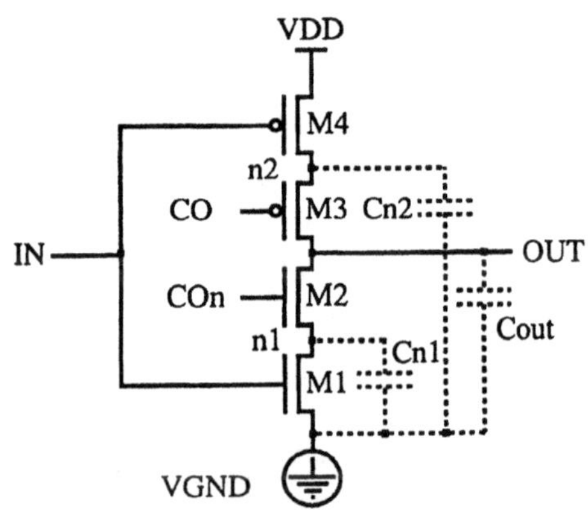

Fig. 14. Tri-state inverter.

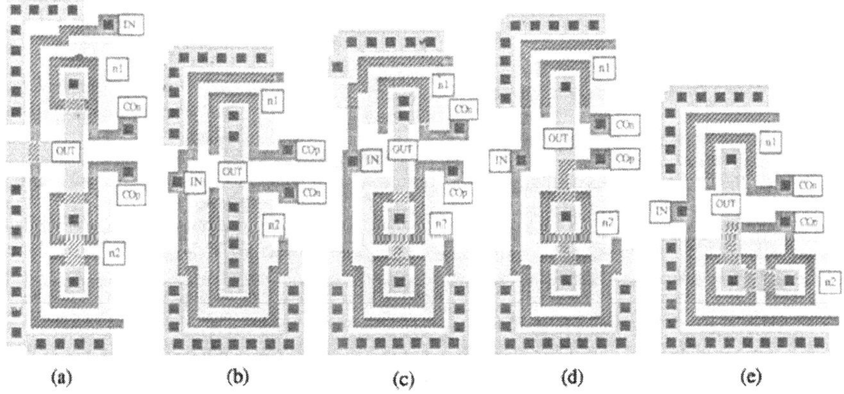

Fig. 15. Different layout implementations of the inverter from Figure 14.

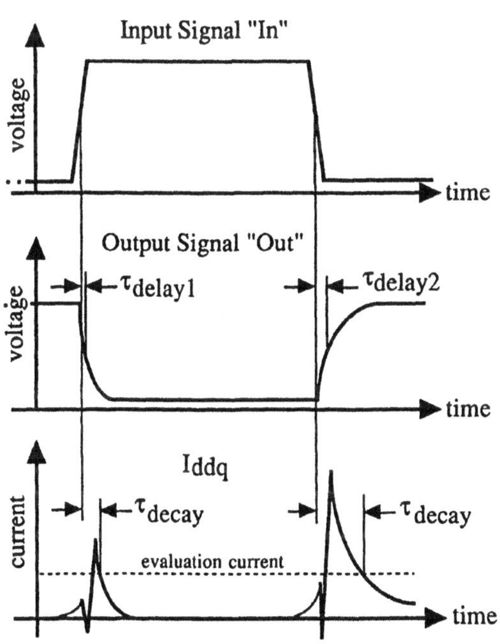

Fig. 16. Definitions of delay and current decay times.

illustrated by the data in Table 1, obtained through a SPICE simulation of the circuits extracted from the layouts shown in Figure 15.

Note that by the decrease of C_{n1} decay2 was decreased by about 30% (obtained for the layout shown in Figure 15e). One can see then that layout optimization is worth the effort since it has not affected circuit performance and has significantly improved the delay/decay ratio.

Table 1. Comparison of decay and delay times.

Layout #	(a)	(b)	(c)	(d)	(e)
C_{OUT} [fF]	31	48	34	34	34
C_{n1} [fF]	49	27	30	25	25
C_{n2} [fF]	69	50	74	74	67
τ_{decay1} [ns]	3.98	3.96	3.97	3.96	3.95
τ_{decay2} [ns]	18.12	13.23	13.62	12.38	12.37
τ_{delay1} [ns]	0.600	0.591	0.641	0.611	0.592
τ_{delay2} [ns]	0.802	0.748	0.725	0.720	0.720

6. Conclusions

In this article the general circuit design strategy for an IC using BIC sensors was described. It was shown that the circuit partition is the key design element affecting both the area overhead needed for the BIC sensor and the testing clock rate. It was also shown that special design rules must be observed during the design in order to avoid abnormally high I_{DDQ} currents.

Hence, BIC testing seems to emerge as a very attractive new BIST technique. But successful applications of this technique cannot be accomplished unless careful design of both BIC sensor itself and the circuit under the test is performed in all stages of IC design process.

Notes

1. Ideally, computation of $I_{DDQ}(N)$ should be done using circuit simulators and from data extracted directly from the layout. But they can also be estimated using the approach illustrated in the examples of Section 3.2.
2. Of course, one can envision usage BIC sensors in CMOS dynamic logic or precharged circuits but in such a case sensors with specially adjusted and carefully designed V_{ref} sources should be used.

References

1. J.M. Acken, "Testing for bridging faults (shorts) in CMOS circuits," *Proc. Dig. Autom. Conf.*, San Francisco, CA, pp. 717-718, July 1983.
2. M. Levi, "CMOS is most testable," *Int. Test Conf.*, Washington, DC, pp. 217-221, September 1987.
3. Y.K. Malaiya and S.Y.H. Su, "A new fault modeling and testing technique for CMOS devices," *Int. Test Conf.*, Philadelphia, PA, pp. 25-34, September 1982.
4. T.M. Storey and W. Maly, "CMOS bridging fault detection," *Proc. of the Int. Test Conf.*, Washington, DC, pp. 842-851, September 1990.
5. W. Maly and M. Patyra, "Built-in current testing," *IEEE Trans. on Solid State Circuits*, vol. 27, pp. 425-428, March 1992.
6. J.M. Soden and C.F. Hawkins, "Test considerations for gate oxide shorts in CMOS ICs," *IEEE Design and Test*, pp. 56-64, August 1986.
7. W. Maly and P. Nigh, "Built-in current testing—feasibility study," *Proc. Int. Conf. on Computer-Aided Design*, Santa Clara, CA, pp. 340-343, November 1988.
8. W. Maly and P. Nigh, "Built-in current testing of integrated circuits," US Pat. No. 5,025,344, June 1991.
9. D. Feltham, P. Nigh, L.R. Carley, and W. Maly, "Current sensing for built-in testing of CMOS circuits," *Proc. Intern. Conf. on Computer Design*, Rye Brook, NY, pp. 454-457, October 1988.
10. K. Kettler, "Current monitoring circuits for fault detection in CMOS systems," M.S. dissertation, Carnegie Mellon University, Pittsburgh, PA, 1989.
11. P. Nigh, "Built-in current testing," Ph.D. dissertation, Carnegie Mellon University, Pittsburgh, PA, 1990.
12. W. Maly and P. Nigh, "Built-in current testing for VLSI circuits," *Proc. of SRC TECHCON'88 Conf.*, Dallas, pp. 149-152, October 1988.
13. W. Maly, "Design methodology for defect tolerant integrated circuits," *Proc. of CICC'88*, Rochester, NY, pp. 27.5.1-27.5.4, May 1988.
14. P. Nigh and W. Maly, "A self-testing ALU using built-in current sensing," *Proc. Custom Integrated Circuits Conf.*, Rochester, NY, pp. 22.1.1-22.1.4, May 1989.
15. D.A. Hodges and H.G. Jackson, "Analysis and design of digital integrated circuits," New York: McGraw-Hill, 1983.

Wojciech Maly recieved the M.Sc. degree in electronic engineering from the Technical University of Warsaw, Poland, in 1970, and the Ph.D. degree from the Institute of Applied Cybernetics, Polish Academy of Sciences, Warsaw, Poland, in 1975. From 1970 to 1973, he was with the Institute of Applied Cybernetics. In 1973, he joined Technical University of Warsaw, where he was appointed Assistant Professor in 1975. From September 1979 to July 1981, he was a Visiting Assistant Professor of Electrical and Computer Engineering at Carnegie Mellon University, Pittsburgh, PA. Since September 1983, he has been with Carnegie Mellon University, where he is a Professor of Electrical and Computer Engineering and Co-director of SEMATECH Center of Excellence.

Dr. Maly is a Fellow of IEEE. He received various awards including SRC Inventor Recognition Awards, Benjamin Richard Teare Teaching Award, AT&T Foundation Award for Excellence in Instructing of Engineering Students and the Best Paper Award from the International Test Conference 1990.

Marek J. Patyra received the M.A.Sc. and Ph.D. degrees from Warsaw University of Technology, Warsaw, Poland in 1978, and 1986 respectively. From 1978 to 1989 he worked as an Assistant Professor at the Institute of Electron Technology, Warsaw. Between 1986 and 1989 he served as Visiting Professor at the Institute of Informatics and Institute of Microelectronics and Optoelectronics, both at Warsaw University of Technology. From 1989 to 1991 he was a Research Associate at the Department of Electrical and Computer Engineering, Carnegie Mellon University, Pittsburgh, PA where he worked techniques for circuit design for manufacturability. Since 1991 he has been an Assitant Professor at the Department of Computer Engineering, University of Minnesota, Duluth, MN. During the summer of 1992 he was a Visiting Professor at the Swiss Federal Institute of Technology, Lausanne. His research interests include VLSI circuit design for built-in and self-testing and VLSI implementation of fuzzy logic circuits.

Subject Index*

* Page numbers in this index are the ones appearing at the bottom of pages.

GPSR Compliance

The European Union's (EU) General Product Safety Regulation (GPSR) is a set of rules that requires consumer products to be safe and our obligations to ensure this.

If you have any concerns about our products, you can contact us on ProductSafety@springernature.com

In case Publisher is established outside the EU, the EU authorized representative is:

Springer Nature Customer Service Center GmbH
Europaplatz 3
69115 Heidelberg, Germany

Batch number: 10615115

Printed by Printforce, the Netherlands